Tl

Nelson Dunkin of Copper Island

by **Jim Badke**

With R. Bruce Scott, Margaret Stewart
and many other contributors

Foreword by Earl Johnson

The Island and I • Nelson Dunkin of Copper Island
by Jim Badke • jimbadke.com

ISBN 978-1-7777101-7-0

Wherever possible, credit has been given for content and photos. Please contact jimbadke@gmail.com to suggest corrections or direct credit to where it is due, for future editions and/or inclusion on the website below.

Front Cover: Source unknown, donated by Debbie Maxie
Back Cover: Photo by Heather Arnott
Title Page: Bamfield Community Museum & Archives

---

nelsonedunkin.blogspot.com

The author will regularly post here stories, photos and other information about Nelson Dunkin as it becomes available.
You will also find purchase and author information here.
Submit additional stories or photos to jimbadke@gmail.com.

Proceeds from the sale of this book will enable kids to attend summer camp who could not otherwise afford to do so. Email the author for more information at jimbadke@gmail.com.

Subscribe to
Jim's newsletter:

Preface

I realized at the start of this project that the best I could hope for was a collection of snapshots of the life of Nelson Dunkin. At first, I was surprised by the limited details I could glean from interviews with people who knew Nelson well. Then I tried accessing my own memories and saw the difficulty. Certain words and images we can remember like yesterday, but they are few. We are left with a general sense of the man and a belief that our friendship with him changed us somehow.

The process of gathering these glimpses together into a book has been satisfying. I especially enjoyed my twenty-plus conversations with people who knew and loved Nelson. Thank you for taking the time. I have met with legends and saints and rascals—I leave it to you to decide your category—and perhaps you fit into all three. There are others I hoped to interview, and if you are one of them, it is not too late. Email me at jimbadke@gmail.com and I would love to talk with you.

Nelson's letters were another treasure trove of information. I am grateful to the several people who shared with me the letters they have kept for so long. Portions of his letters are quoted in this book; for the sake of the recipients' privacy, I avoided including especially personal bits. As I transcribed the letters, I preserved as much as possible of Nelson's figure of speech; however, I corrected some spelling and grammar where necessary for clarity. I have used Canadian spelling throughout, which Nelson adopted as his own.

Most of the photos scattered throughout the book have no credit attached to them, as I could not always verify the identity of the photographers. If you recognize one of your photos and wish to have credit, please contact me and let me know (jimbadke@gmail.com). I will include your name in future editions and the nelsonedunkin.blogspot.com website.

Thank you to those who gave their feedback on the manuscript: Madge Vallee, Nelson Dunkin II, Earl Johnson, Dave & Inger Logelin, Aaron Otis, Heather Arnott, Debbie Maxie, Michele Sadler and my wife, Sarah. Your suggestions and corrections have fine-tuned the book far better than I could have accomplished on my own. Thank you also to Margaret Stewart for your chapters, Earl Johnson for your gracious Foreword and Susan M. Scott for permitting me to include the chapter from your father, R. Bruce Scott.

The purpose of this biography is to honour the memory of Nelson Dunkin and give thanks to the God he loved and served. I haven't hesitated to include—as one interviewee put it—the good, the bad and the ugly, not to mention the beautiful. But I hope that in everything, the reader will observe the grace of God upon which Nelson so heavily depended. May you acquaint yourself with Nelson, relive the memories you may have of him, and come to a greater understanding of what made him the man he was.

- Jim Badke, Honeymoon Bay, BC, November 2023

Photo: Bamfield Community Museum and Archives

Contributors

Many people contributed to this project, telling me their memories and stories and showing me photos, letters, artwork and carvings. Together, we shared our affection for and memories of Nelson and Mina Dunkin. The following pages will mention many of these people, and I will usually refer to them by their first name:

- **Aaron Otis**: Current Director of Ministries at Copper Island Camp, involved there since 2002.
- **Ben and Irene Potter**: Friends who frequently fished out of Nelson's place.
- **Bill and Laurel Irving**: Friends of Nelson; Bill was the former mayor of Ucluelet.
- **Bill Priest**: Frequent visitor who had a fish farm across the channel.
- **Brian Burkholder**: Coastal Missions missionary and friend from their earliest days on Copper Island.
- **Dave and Inger Logelin**: First directors of Copper Island Camp (Wilderness Retreat Society).
- **Debbie and Tom Maxie**: Coastal Missions missionaries who visited often; Debbie trained at Camp Ross.
- **Don McNaughton**: Shantymen missionary who visited Nelson by float plane.
- **Donna Green**: Visitor, along with her father and uncle who were logging on Copper Island.
- **Earl Johnson**: Shantymen missionary on MV *Messenger III* and still going strong at 94.
- **Harold Sadler**: Regular visitor with his dad Jim, who was a key builder of Nelson's new house.
- **Heather Arnott**: Trainee at Camp Ross who often corresponded with Nelson.

- **Heather Cooper**: Bamfield Museum archivist who provided historical photos and contacts.
- **Jim and Sarah Badke**: Author/editor and wife, brave enough to bring an 8-month-old baby to Nelson's.
- **Joan (Petunia) Getman**: Coastal Missions missionary who directed Camp Ross.
- **Len Gedak:** Artist who visited Nelson and did paintings and photography there.
- **Leona Dolling**: Nelson and Mina's granddaughter, daughter of Madge.
- **Lynn Starter**: Visitor who provided a painting of Nelson's house.
- **Llyod Bridal**: Neighbour from the Port Albion days, as interviewed recently by Phil Hood.
- **Madge Vallee**: Daughter who lived on Copper Island for about 5 years before starting school.
- **Margaret Stewart**: Mina's relative by marriage who lives on the Isle of Lewis in Scotland, where Mina grew up.
- **Nelson Dunkin II**: Son who lived on Copper Island for about 3 years before starting school.
- **Pat Rafuse (Lovett)**: Friend and occasional visitor who provided several letters.
- **Patty Cameron**: Friend who, with husband Don, lived with Nelson for two years on Copper Island.
- **Peter Horton**: Neighbour who still lives in Kildonan, from where he often visited and assisted Nelson.
- **Phil Hood**: *Westerly News* editor who knew Nelson and Mina from the Port Albion days.
- **R. Bruce Scott**: Historian who included two chapters about Nelson and Mina in one of his books.
- **Rich Parlee**: Pastor in Tofino and Ucluelet who visited regularly.

- **Rick Charles**: Artist and musician who visited, played violin and sketched several of the drawings in the book.
- **Ron Pollock**: Friend who contributes posthumously through the collected letters Nelson wrote to him.
- **Roy Getman**: Coastal Missions missionary and boat builder who visited from the early days.
- **Ruth Sadler**: Wife of Jim Sadler, who was instrumental in building Nelson's new house.
- **Steve Priest**: Occasional visitor who had a fish farm across the Channel.
- **Susan M. Scott**: Visitor with her father (R. Bruce Scott) who provided permission to include his story.
- **Wendall Ferrell**: Brother to Patty who visited often.

Photo: Aaron Denninger

Contents

The new house in 1986: shop below, living area above; built by friends, Nelson style.

Foreword

by Earl Johnson

It was June 1952 and my first week aboard MV *Messenger III* as a 22-year-old shipmate to a seasoned 55-year-old skipper, Harold Peters. We tied up alongside the Dunkin family float house in Kildonan, Barkley Sound. I have a crystal-clear visual memory that still warms my heart of two beautiful, quiet children standing beside their parents. Nelson Jr. was six; Madge was four. Little did we know that both of these children would one day live with us while attending high school in Port Alberni.

Today, 72 years later, I'm honoured by the invitation to write a Foreword to Jim Badke's comprehensive biography of Nelson Dunkin, a hidden-away man among men.

Many individuals on Vancouver Island's West Coast choose utter seclusion in secret coves. Nelson, quiet and humble by nature, also chose to live his gifted life in nature's relatively private beauty. But his life was not so secluded, as you are about to find out!

The fishing industry in which Nelson once worked saw significant changes in his lifetime. Salmon canneries were closing as it became more profitable to transport fish by refrigerated packers to Steveston, outside of Vancouver. The wartime deportation of 5000 Japanese fishermen and their families out of Barkley and Clayoquot Sounds in March of 1942 devastated the area. The Japanese were then the largest of the three ethnic groups in that area. Ten years later, in the mid-1940s, Skipper and I were privileged to enjoy good fellowship with the Japanese families that returned to Ucluelet as well as with many of the First Nations and Caucasian families.

The rough Pacific Ocean side of Vancouver Island—from Race Rocks Lighthouse in the south to Cape Scott Lighthouse at the top end of Vancouver Island—was our assignment with the Shantymen's Christian Association. Surely, a quiet evening with the Dunkin family was among the best. The opportunity to honour the legacy of the Dunkins means a lot to me.

As a veteran of World War II, Nelson rarely spoke. The term used back then was "shell shocked." Today, we would say he suffered from post-traumatic stress disorder due to the war. Healing came to Nelson through:

- Living in isolation
- Marrying a woman with a beautiful Scottish brogue whose voice brought great delight to all who heard it
- Raising two engaging children
- Carving words and prayers with beauty and intention throughout his home
- Constructing a house with dreams of hospitality for years to come, welcoming both friends and strangers alike

As it is said, Nelson preached Christ always, and only when necessary did he use words.

Barkley Sound Vacation Bible School Teams in 1954, Earl on the left.

My children's fondest memories of Copper Island are:

- Learning to walk and then run along the logs tethered together as a path from the boat to the beach
- The innovative way Nelson and Mina recycled their cans
- The bright red yolks from their free-range beach chickens
- The huge pancakes Mina made on the wood stove, and her engaging games of checkers
- The two-storey cabinet with creative carvings that depicted from Revelation the horrors of hell and the joys that await us in heaven
- Nelson and Mina's warm smiles and hugs
- The freedom to explore the beach and the outdoors
- Listening to the din of the late-night conversations while tucked into cozy beds adorned with carved prayers—and decorated with inlaid marbles—to be read and repeated

I hope you enjoy this book and are reminded of the seeds that Nelson and Mina planted well in the lives of all whom they encountered in the remoteness of Copper Island. Each visit with them left me and my family or crew feeling encouraged and well-fed. Our goodbyes were always bittersweet as we loosened the ropes from their float, singing, "God be with you 'til we meet again."

Mina and Nelson never lived to see the fullness of the fruit of the harvest. Nonetheless, they left us the gift of seeds sown, and here we are, reaping the benefits.

May you be encouraged to sow seeds well in your life, not only for today but also for generations to come. This was the vision and dream of Nelson and Mina, and the legacy they left to carry on in Jesus' name.

- Earl Johnson, October 2023, Campbell River

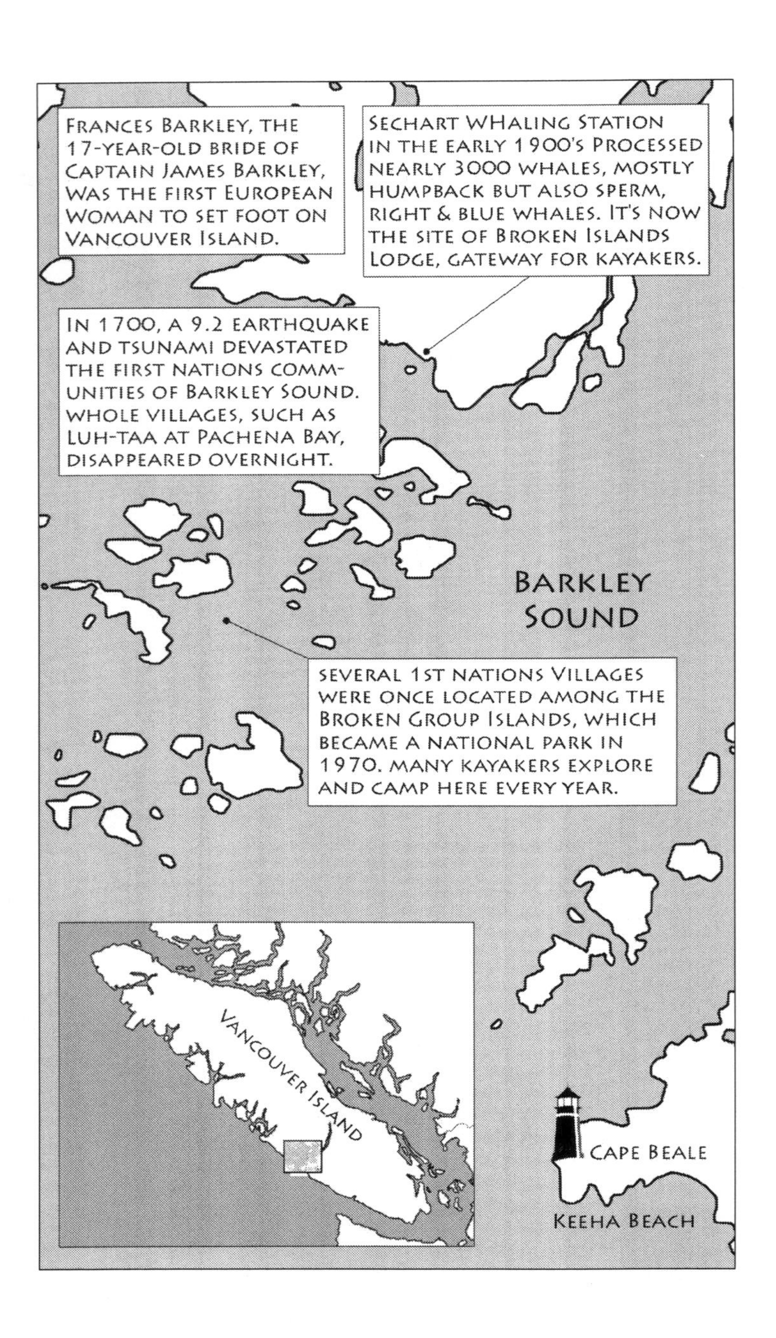

Frances Barkley, the 17-year-old bride of Captain James Barkley, was the first European woman to set foot on Vancouver Island.
Sechart WHaling Station in the early 1900's Processed nearly 3000 whales, mostly humpback but also sperm, right & blue whales. It's now the site of Broken Islands Lodge, gateway for kayakers.
In 1700, a 9.2 earthquake and tsunami devastated the First Nations communities of Barkley Sound. whole villages, such as Luh-taa at Pachena Bay, disappeared overnight.
Barkley Sound
several 1st Nations Villages were once located among the Broken Group Islands, which became a national park in 1970. many kayakers explore and camp here every year.
Vancouver Island
Cape Beale
Keeha Beach

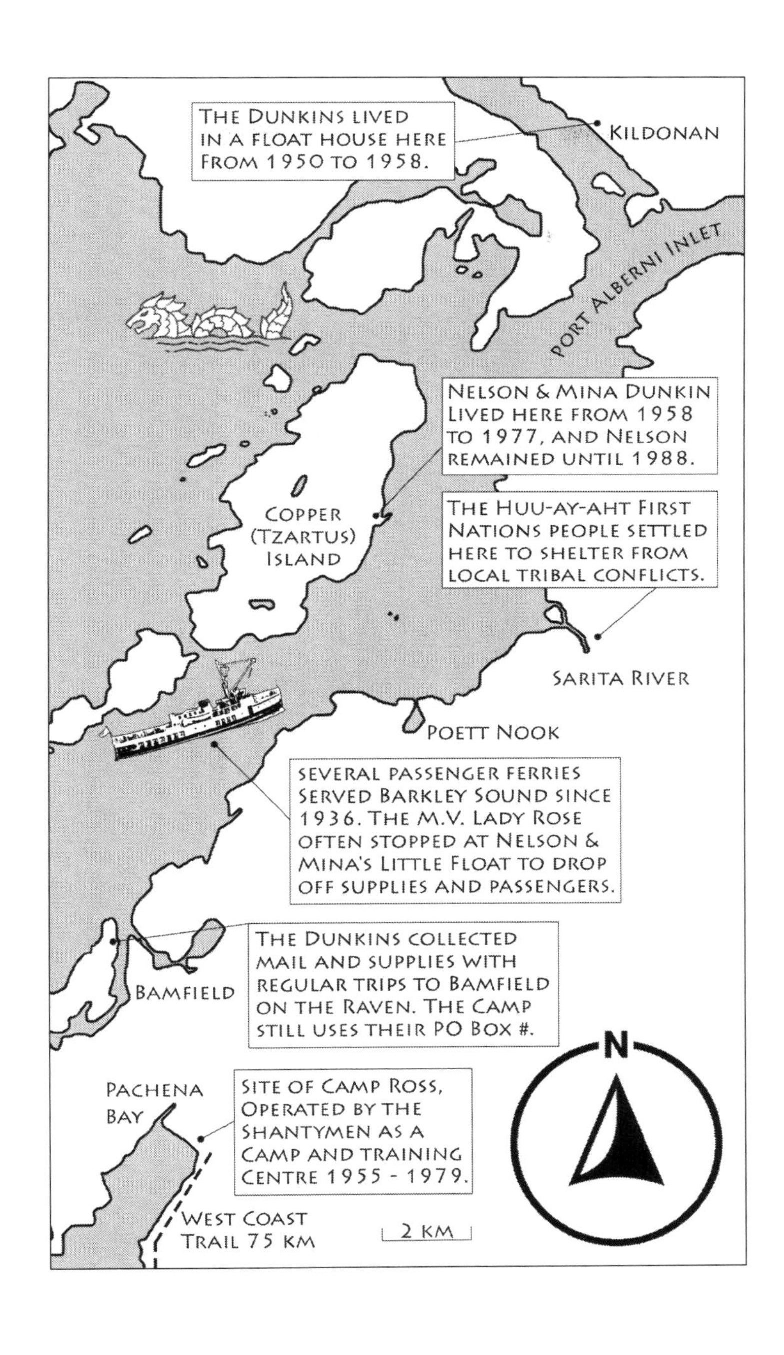

The Dunkins lived in a float house here from 1950 to 1958.
Kildonan
Port Alberni Inlet
Nelson & Mina Dunkin lived here from 1958 to 1977, and Nelson remained until 1988.
Copper (Tzartus) Island
The Huu-ay-aht First Nations people settled here to shelter from local tribal conflicts.
Sarita River
Poett Nook
Several passenger ferries served Barkley Sound since 1936. The M.V. Lady Rose often stopped at Nelson & Mina's Little Float to drop off supplies and passengers.
The Dunkins collected mail and supplies with regular trips to Bamfield on the Raven. The Camp still uses their PO Box #.
Bamfield
N
Pachena Bay
Site of Camp Ross, operated by the Shantymen as a camp and training centre 1955 - 1979.
West Coast Trail 75 km
2 km

Nelson's dock and log boom, Copper Island.
Pen and ink drawing by Rick Charles, 1984.

1. The Lower Case i

When I (Jim) was 19, I travelled from my home in Kelowna to a summer job with Camp Ross at Pachena Bay. It was my first time on the west coast of Vancouver Island. The constant rumble of surf, the impenetrable forest and the sculpted headlands fascinated me. My primary role was to inform and register hikers at the north end of the fabulous 75-km West Coast Trail, which was part of a new national park.

Many interesting people passed through the A-frame trail hut that summer of 1978. I met one fellow who had hiked the trail packing nothing but a trombone in a case. Camp Ross saw my work at the trailhead as a way to come alongside people and leave them with something of the care of God.[1] The camp was a training centre for young missionaries and ran a full summer program of kids' camps and youth retreats.

For one of those retreats, the camp asked me to come along as a counsellor. This would be an overnight trip to an island in Barkley Sound. We all boarded Brian's fishboat, the *Kolberg*, and motored out of Bamfield Inlet and 10 km up Trevor Channel. As we rounded a point and entered a small bay, I was surprised to see a large cedar-clad house, three storeys tall, with an extra lookout on top. A white-haired gentleman, followed by a black dog the size of a bear, was navigating a series of logs to greet us. His tiny float was dwarfed by the fishboat, but he caught the lines tossed to him and deftly moored us.

This was my introduction to Nelson Dunkin's home on Copper Island. My proper introduction to the man himself came the next day. Nelson had built his house on pilings over the shore, which meant that the high tide came up under the house twice daily. "I wouldn't build it the same way again," he told a reporter many years later. "When you have a house on posts, you've always got to watch that logs don't get underneath and pry the thing apart."[2] To avoid this, he had made a breakwater of upright logs along the bottom of the house, facing the water.

One of the many ways we helped Nelson that week was to continue the breakwater along the bridge that ran from house to shore. They put me in charge of the project. Being new to coastal construction, I was nervous about my responsibility, but I gamely led the way. We gathered driftwood and cut it to length. Then below the ramp, we dug down a couple of feet in the sand and gravel of the beach. Finally, we set our uprights in place, nailed and wired them to the ramp above and filled in the holes.

At one point as I was doing this, I became aware that the white-haired gentleman was leaning on the ramp above me and watching intently. I greatly lacked self-confidence at that age, and my apprehension grew, thinking that perhaps I was getting it all wrong. Finally, I stopped and glanced up at him. Nelson looked at me squarely and said, "How did you know how to do that? You did that exactly right!"

This was the beginning of a friendship that lasted twenty years, and the initiation into a way of seeing that has lasted a lifetime. My story has been profoundly shaped by the life and example of this godly man, Nelson Dunkin. Because of this, I have always felt I should attempt to tell Nelson's story. Over the past few months of interviews and research, I often wished I had started sooner.

Nelson went home to be with Jesus some 25 years ago, and most of Nelson's closest friends have now joined him. Pictures have faded; letters, journals and memories have been lost along the way. However, Nelson made a profound impression on many people. From them, I have gathered enough material to assemble a fair picture of his life, his work and his legacy.

Everyone's eagerness to contribute to his story amazed me. They were the source of many memories—and memorabilia—of Nelson. If you are one of these people, I thank you and hope I have represented you well. Nelson's daughter Madge commented on the many letters from her father that people have shown her. "I mean, these people kept these letters because I guess they were so interesting." That is to say, her father was an interesting and intriguing character.

Debbie asked me, "So if Nelson knew you were writing a book about him, what would he say?"

I laughed and replied, "Yeah, there would be a few extra twitches in his face, and he wouldn't be very excited about it." After a moment's reflection, I continued, "He would want me to talk about all the people who came to see him—everyone who was involved in his life—and not just talk about him."

As I was typing out his many letters, I realized how true this was. Not only did Nelson mostly write about other people, but every time he came to the word "I" in his letters, he did not capitalize it. As it takes more effort to write "i" than "I" (and modern spellcheckers make it harder to type), we can assume that his lower case i's were intentional. He was saying all the time, "This is not about me."

So… my apologies, Nelson—this one is about you.

And yet in some ways, it's not. This book is also about the people who knew him and the God who knew him best. Peter of Kildonan told me, "Some people are overly religious and pounding the Bible in your face all the time, but Nelson was not that guy, not at all. He lived it and led by example, which gave us a lot of momentum for our lives down here." Nelson would be happiest if, by reading this book, some people might follow his example as he tried to follow the example of Jesus.

Actually i have the patience of Job and even so, "Sometimes i'm up and sometimes i'm down." But it is to be doing God's will. So i continue to struggle onward.[3]

[4]

---

[1] From an interview with Earl Johnson, April 25, 2023

[2] Katie Poole, "An Island of Solitude," *Tri-City News*, November 17, 1991

[3] Letter to Jim and Sarah Badke, October 25, 1986

[4] Bamfield Community Museum and Archives

2. First Impressions

I often asked contributors in interviews, "What was your first impression of Nelson Dunkin?" A common memory was the twinkle in Nelson's eye, which is well captured in the cover photo.

When I first arrived at Copper Island, it was with several people whom Nelson knew well. I remember wanting to be part of that comradery. He looked a bit alarming with his wild hair and pronounced features. I wondered what he thought of me, if he thought of me at all. But I found myself drawn to Nelson in those few days, and the memories of that visit stuck with me. I was determined to return to Copper Island and get to know this man better. It would be five years before I had the opportunity.

Here are a few other first impressions of Nelson Dunkin:

- **Debbie Maxie:** He had that sparkle, and he was always a bit hunched over. He had a twitch, which I guess was left over from the war, shell shock or something like that.
- **Ruth Sadler:** Nelson was quiet, had a sense of humour and was very intelligent and good with his hands. He was always inventing things and making things work, getting along without electricity. Nelson was artistic and creative, as seen in the beautiful cradle he made for his granddaughter. He was well-loved by everyone.
- **Brian Burkholder:** Nelson and Mina were always, always very kind to anybody who came to their place. I guess that's why I kept going back.

- **Inger Logelin:** My first impressions of Nelson were his sprightly manner, his twinkling eyes, his intelligence and humour, and his humility.
- **Earl Johnson:** Nelson was meek, wonderful, gracious. A quiet soul, but very knowledgeable in life. I don't know if I'd call him a recluse. I visited the recluses on the coast.
- **Heather Arnott:** I remember he had a tremor and a shock of white hair, and always a kindly half-smile on his face. He looked older than he probably was.
- **Dave Logelin:** Nelson and his wife were a real lighthouse to the community around them. They encouraged many people through their prayers, their witness—especially Mina—and their hospitality. Nelson was his own person. He reminded me of Einstein with his full white head of hair. He wore a pair of cut-off logger pants held up by one strap (it may have been a bungee cord). In conversation with him, I found he was a private, independent person. He didn't think too highly of the government and didn't want any interference from them. Nelson had a subtle sense of humour.
- **Harold Sadler:** Nelson was kind and soft-spoken. But he was hard to get to know, a bit intimidating in his looks.
- **Peter Horton:** As far as we're concerned, Nelson was a pretty straight shooter. He lived his life. He didn't proselytize. He was a Christian, but he just lived it; he didn't have to talk about it.
- **Margaret Stewart:** He was a lovely man. A lovely, gentle man. He was very special. I imagined it was just perfect, living on that little island. I know it was difficult at times.
- **Bill Priest:** He'd come tottering down the ramp with all the floats and the planks moving, just as nimble as could be, to smile, to help tie up the boat and say, "Come on up for a cup of tea."

- **Pat Rafuse:** I remember he just had a way about him. He would putter around at a certain pace, and he was never a person to appear busy or under pressure to get anything done. But it was amazing what he did.
- **Ben Potter:** Beneath this less-than-impressive face and stature, there lurked a powerful and creative brain.
- **Patty Cameron:** He was so thin and sad. His calendar was marked with the days since his wife passed away. We decided to stay for a while and ended up staying for a couple of years.
- **Bill Irving:** Nelson and Mina were a pretty vivid example of—not only unique characters on the West Coast, novel individuals who made a home for themselves out in the middle of nowhere—but also faithfulness and trust that the Lord was providing.
- **Adele Wickett:** He was an individualist, for sure. He was one of those characters who are drawn by the wild west coast of BC, and stay to contribute their own stories to its culture. And, in this he was a Christian believer who wove his life into what God was doing there.[1]
- **R. Bruce Scott:** [Nelson and Mina] are not hermits in any sense of the word, being very sociable, but they like the independent, and consequently isolated, way of life.[2]
- **Don McNaughton:** My most enduring memory of Nelson was his complete and unhurried contentment to abide in Christ, waiting patiently for opportunities to share with others. From those few visits with Nelson Dunkin, I was permanently changed.
- **Ron Pollock:** I just really *love* Nelson.[3]

---

[1] Adele Wickett, "Remembering Nelson Dunkin," *Island Christian Info*, June 1998.

[2] R. Bruce Scott, *People of the Southwest Coast of Vancouver Island*, 1974

[3] From Heather Arnott's memories of Ron.

Don Cameron had Nelson's brass Psalm 23 plaque fixed beside the captain's chair on the boat he rebuilt, the *Nelson E Dunkin.*

3. Beginnings

How did Nelson Dunkin become the person who made such an impression on so many people? We need to go back to the beginning of his story. And yes, there will be some dates, names and places here. But if you stick with me, you will also hear stories about Nelson you may never have heard before. Nelson didn't always live on an island, and some of his background will help make sense of what you might already know about him. Many of the following stories were told to his children or distilled from historical records.

He was born Nelson Edward Dunkin on January 22, 1909, in Olympia, Thurston County, Washington, USA to Jonathan Cook Dunkin (from Kansas) and Rhoda May Dunkin (nee: Tedhams, from Michigan), who married on June 6th, 1884.[1]

Nelson's father was a farmer, carpenter and later a logger who dragged the family around to various logging camps in Washington, Oregon and Idaho. Nelson II remembers his father talking about "how much they had been here and there." Photos show that Nelson's family often lived in a canvas wall tent. His Irish-heritage father John was also a fiddler in various Irish bands. The combination of logging camps and pubs lent itself to a wild lifestyle. "Alcoholic, gambler, not a good person," was what Nelson's daughter Madge remembers hearing. "I forgot to throw in womanizer too. My dad didn't have much use for his dad." But Nelson was fond of his mother. On top of shepherding six children in their father's tracks, his mother Rhoda was the cook for a road construction gang.

Nelson and pup. Long Lake was the home of the infamous meeting place, the Buckhorn Café, where Nelson's father undoubtedly played many an Irish jig.

Nelson's siblings at the time of his arrival included brothers John (12), Clyde (10), Harvey (8), and sister Avis Ruth (7). There was also an older sister named Madge who passed away as a baby. Nelson did not have a good childhood or family life. "His family was dysfunctional before 'dysfunctional' became a word for families," said daughter Madge, who was named after her baby aunt. Despite their nomadic life, census records indicate that the family gravitated around Thurston County, southwest of Seattle, near Olympia where he was born.[2]

The Dunkin family. Nelson is centre bottom. Baby Madge had already passed away.

From the photos, it appears that at about age five, Nelson had a dog named Buck. Later, he also had a pet pig named Lilly that he raised from piglet to adult. Nelson often told a story from when he was four or five. He was out riding a horse on a stormy day when lightning struck! As it did, he saw an angel pluck him off the horse and set him on the ground. The lightning killed the horse but spared Nelson's life.

Nelson's elementary education was hit and miss. Many years later, he wrote to his granddaughter Leona, "i was at school but couldn't learn nothing."[3] Nelson II told me that, in the second grade, the teacher put a dunce cap on his father. When Nelson went home and told his dad about being humiliated in front of his class, his dad replied, "Well, you don't have to go back to school then." His dad kept him out of school for two full years. Somehow, he completed grade eight at St. Michael Parish School in Olympia, which was as far as the school—and his childhood education—went.

Young Nelson with Lilly, his pet pig that he raised to full size.

When he was 15 years old, Nelson's fragile world fell apart. His mother Rhoda passed away at the young age of 47, just a few years after his favourite sister Ruth had passed away at 19. As the only minor left in the home, Nelson was sent off to live with his Aunt Emma Hawley on a small farm in the Woodard Valley near Olympia. Nelson II remembers that his father showed him where the house once stood—right under the current Interstate Highway 5. This home was not a significantly better environment, but Nelson lived there until he was at least 21. Census records note that besides his aunt, the household included her elderly mother and a male "lodger," none of whom were employed.

But at least the farm was teaching him to work hard, among other new abilities. Patty remembers Nelson's delight in telling stories about the family "still." These were the days of American Prohibition, when the Temperance Movement led to the legal prevention of the production, sale and transportation of alcoholic beverages. Nelson would take a bottle of family moonshine and bury it in the field where he was working, always next to a dandelion. He would take a dandelion stem and set it in the open bottle. "So on hot days, he'd be working hard, and then he'd go back to his little dandelion and take a sip," Patty relates. "We laughed so hard, and we thought, Oh, Nelson!"

No one seems to know how Nelson came to faith in Jesus. Nelson II understood that his father was a believer from his youth but didn't know the circumstances. He believed his father had attended a Presbyterian church at some point, possibly while living with his aunt. In any case, as Nelson's story unfolds, it becomes clear that his faith in God was central to his way of life. The same was true of his wife Mina, who grew up regularly attending the Free Church of Scotland. "Both mom and dad were strong in faith and loved the Lord," Nelson II recalls. "They went through different phases, but were very constant in their walk with God."

At some point, a man in Olympia took in Nelson to train him as a machinist. In the late 1920s, American families were buying cars, washing machines and refrigerators on credit for the first time, leading to a boom in production. Many of the interests and skills that would carry Nelson through life on Copper Island began here. "Living on the Olympic Peninsula, Dunkin watched Irish Catholic blacksmiths and learned enough to run his own smithy."[4] He also gained skills as a locksmith and learned to create his own locking systems. While working at the machine shop, Nelson decided he wanted to go to college. He applied and was told that if he could pass an entrance exam, he would be upgraded. He was accepted into the college, but it is unclear how long he remained there.

The next mention of Nelson is at the age of 31. This was at the end of the Great Depression, and Nelson had been unemployed for 32 weeks that year. He was living on a farm in Gull Harbour near Olympia and had found a job as a Timekeeper for the Work Projects Administration. This was a government make-work program that "employed mostly unskilled men to carry out public works infrastructure projects. They built more than 4,000 new school buildings, erected 130 new hospitals, laid roughly 9,000 miles of storm drains and sewer lines, built 29,000 new bridges, constructed 150 new airfields, paved or repaired 280,000 miles of roads and planted 24 million trees to alleviate loss of topsoil during the Dust Bowl."[5] The Timekeeper kept track of who had worked and made out the payroll so cheques could be issued.

At this time, Adolf Hitler invaded Poland, driving the United Kingdom, France and the British Commonwealth to declare war on Germany, soon to become World War II. Nelson, barely employed and having a falling-out with the household, decided he wanted to go and be part of the action. However, the United States of America had not yet joined the Allies. In late 1940 or early 1941, Nelson crossed the border to join the Canadian Army in Vancouver. This decision would change the course and fortune of his entire life.

---

[1] "Nelson Edward Dunkin," myheritage.com

[2] "Census of the United States, Population Schedule," 1910, 1920, 1930, myheritage.com

[3] Letter to Leona Dolling, January 17, 1972

[4] Katie Poole, "An Island of Solitude," *Tri-City News*, November 17, 1991

[5] history.com/topics/great-depression/works-progress-administration

4. Ingenuity

When Nelson crossed into British Columbia and enlisted in the Canadian Army, he was immediately sent for training at the Bay Street Armoury in Victoria. This was his first visit to Vancouver Island. Later, he was sent to Cold Lake (Alberta) and Quebec. With the increase in mechanized war equipment, people with specific trades were needed to produce and maintain weapons and vehicles. His background and interest in machinery qualified Nelson to work on military equipment in the Canadian Forestry Corps.

All his life, Nelson's ingenuity enabled him to develop whatever was needed for the job. Madge said her father "could be considered a genius; he could invent whatever." But in humble Nelson fashion, he seldom received recognition for his contributions. While serving in Quebec, Nelson invented a device that made it possible to release army tanks from trucks faster than ever before. His invention was used by the Allied Forces throughout World War II. Nelson II remembers seeing the drawings of his device that were unfortunately lost in the fire (more on this to come). His contributions to the war effort are lost to history.

Also lost was some remarkable correspondence with great minds around the world. Many people have commented that Nelson's appearance reminded them of Albert Einstein with his wild white hair, large nose and wrinkles. What many don't know is that Nelson actually exchanged letters with Einstein. Nelson was interested in theories of energy and perpetual motion. Einstein stated that the Second Law of Thermodynamics—that energy cannot be created or destroyed—was one that would never be overthrown, and of course, people have been trying to do so ever since. Nelson wrote to Albert Einstein with his ideas, and Einstein wrote him back. Nelson II remembers that Einstein was impressed with his father's theories and concepts. They sent three or four letters to one another, which again were lost in the fire. They would be very valuable today.

He also communicated with other theorists, for example, about harnessing the wind to propel cargo ships. Rather than conventional sails, this concept used fixed turbines that would rotate in the wind and create some form of propulsion—either directly or through electricity. He had many drawings and diagrams for these systems. Among the few things Nelson gave me was a small wooden model with two interlocking discs, angled perfectly to spin rapidly in a breeze from any direction. Nelson would be interested to know of current projects to power cargo ships by this very method, with huge towers that turn in the wind.[1] [2]

But most of Nelson's ingenuity was practical, not theoretical. On Copper Island, he eventually gave up on electric generators, and his ideas for a waterwheel at the waterfall on his property were still out of reach. So Nelson had to create tools and machines for his work that were manually operated. He wrote, "i have sent for a book on the design and building of machines, 2500 pages for $78.00, so it should have a few ideas and hints."[3] Dave remembers, "Nelson was an inventor and could make anything out of very little and make it work. He showed me his pedal-powered wood lathe and grinder, and a heavy-duty hand-cranked drill press that worked well. I didn't know that day I would be using those tools for many years to come."

Nelson was quick to supply other people with the equipment they needed but couldn't find. Brian remembers, "Nelson made me an extra-long hook for holding open the dutch door of my big fishboat, and it was perfect for what I needed. I didn't even ask; he just made it and gave it to me one time when I visited. He was so handy at making things you couldn't buy." Peter still has a set of callipers that "Nelson made all by hand, all ripped together. He was quite the guy. And on his Alaska mill—because the oilers that lubricate the chain are bulky—he rigged up a tank at the end of the bar that had a drip feed onto the chain, which was ingenious. That kept the chain lubricated perfectly."

Rick remembers a game that Nelson created, based on John Bunyan's *Pilgrim's Progress*. It was a tabletop game almost the size of a tabletop, like a large tray with a four-inch rim around it. Inside was an ingenious obstacle course with carved wooden objects related to the parts of the book. Players had a ball or marble that they had to navigate over or under or through or around the obstacles. No one quite remembers how the ball was moved. Possibly, players had to manoeuvre the box itself to get the marble rolling in the required direction. The Sadler family had the game for years but lost track of it after moving several times.

Finding the materials to match Nelson's designs was his constant frustration on the island. He wrote to Don and Patty, "The lathe i was working too long on is at last finished and painted: now comes the making of the attachments. Those belts came just in time as i was ready to measure the length and send to Ostrom's to buy one. Also i happened to have a belt coupler and that worked out just right (first one i ever used). Clever device."

Gears, flywheels and belts: Nelson's foot-powered wood lathe in the new house.[4]

But he added a copy of a note he had sent to one of his suppliers:

> [Company Name] Used Truck Sales, Tofino
> Dear Sirs (or whatever):
>
> Your unsolicited parcel arrived by plane some days ago and i was so very angry that i started biting the contents but when i came to the "V" belts i had time to reconsider and decided not to return the parcel.[5]

---

[1] "Can Massive Cargo Ships Use Wind to Go Green?" nytimes.com/2021/06/24/magazine/cargo-ships-emissions.html

[2] Photo by Aaron Otis. This model is displayed in the Heritage Room

[3] Letter to Jim and Sarah Badke, November 1988

[4] Bamfield Community Museum and Archives

[5] Letter to Don and Patty Cameron, January 3, probably 1981

5. On the Edge of Two Oceans

by Margaret Stewart

Margaret Stewart was born in 1956 and brought up in the Isle of Lewis, the largest and northernmost of Scotland's Outer Hebridean chain, which shields the west coast of mainland Scotland from the worst ravages of the Atlantic Ocean. It is separated from mainland Scotland by the Minch, a rather hostile channel of water around thirty miles wide in some places. It was on this island that Mina Dunkin was born and raised, and it was the place she always referred to as home.

The Callanish Stones bear witness to the occupation of the Isle of Lewis at least 6,000 years ago, though who these early settlers were is unsure. Photo: Margaret.

The people of the Western Isles are a mixture of Celt and Norse heritage. From the 13th to the 16th centuries, the island was variously ruled by the Norse, the Lordship of the Isles and the MacLeods of Lewis. It was then forfeited to the Scottish

Crown, who agreed to allow a group of nobles to colonise and anglicise Lewis and exploit its natural resources. But the ruling MacLeod clan forced them to abandon their plans. The island was later ruled by the Clan MacKenzie of Kintail until it was bought by Sir James Matheson in 1844 and finally, in 1918, by the industrialist Lord Leverhulme.

Try as the ruling classes might, they didn't completely manage to succeed in assimilating us and destroying our language and culture. Scottish Gaelic is still my mother tongue! Like most of my contemporaries, I didn't learn English until I went to school. It would certainly have been the same for Mina. However, many of our place names, coastal features, nautical terms and bird and fish names are borrowed from Old Norse and have been integrated into Scottish Gaelic over time. Coll, in the district of Back, was the village where both Mina and I were born and brought up. We are related by the marriage of her aunt and my great-uncle.

Mina was born on 9th March 1914 to Mary and Murdo MacIver at the family home on Croft number 31 in the village of Coll. She was called "Seumasinag" after her maternal grandfather, Seumas, of number 34 Coll, but she was registered by the anglicised name "Jamesina," as registrars encouraged parents to record only an officially recognised English version of a Gaelic name. Mina was known by the Gaelic version of her name, along with her patronymic, Seumasinag Mhurchaidh Ailean (Seumasinag, daughter of Murdo, son of Allan), as was the custom. Later in life, Seumasinag's name was shortened to Mina, although I am not sure if this was when she went to Canada or when she was working in Edinburgh.

Mina's life on Lewis would have been similar to my own upbringing in Upper Coll. We lived in very close-knit crofting communities. A croft is a small portion of land that can be described as a very small farm, consisting of a few acres. Many of these crofters were also fishermen. All members of the family—young and old—helped to sustain a livelihood on these small plots of land, planting potatoes, root vegetables, cereal crops, and growing hay to feed their cattle.

Crofters were not permitted to subdivide a croft between family members. With growing families and the need for more crofting land as the children grew up, overcrowding became a problem. Emigration was encouraged as a means to relieve this so-called "congestion." This also served the plans of the British government to retain and defend its overseas colonies by settling them with white Protestant settlers.

Lewis is mostly covered in a thick mantle of blanket bog. Though now a conserved wetland, for hundreds of years these peatlands provided the people of the island with their precious fuel. Each household was allocated a peat-cutting area that each year they cut and dried, carted to an area close to their homes and piled into peat stacks. The work was started in spring by removing the uppermost turf, and the task of digging the peat was undertaken communally, with groups of neighbours working together. This was also the custom for many other tasks, such as sheepshearing, potato planting and harvesting.

The moorland also provided vital grazing for the people's cattle during the summer months, following the scarcity of green grazing areas during the harsh Lewis winter. From each household, young mothers and their children—or young milkmaids—spent the summer out on the moorlands, living in small settlements of temporary stone-built, one-room huts, lined with turf on the outside. They stayed at these "shielings" from May to August, milking the cattle and making cheese and butter, which was taken back to the home village regularly. The menfolk tended to the crops, undertook vital maintenance work in the coastal villages or went off to work at the seasonal herring fishing.

During Mina's childhood, most people in the villages of Lewis still lived in what is called a 'blackhouse': a dry-stone-built dwelling with double walls, packed with earth between and roofed with wooden rafters, which were then thatched with straw and held down with netting and heavy stones. None of these remain as dwellings now, but a blackhouse is preserved in the village of Arnol, as well as a village at Na Gearranan, Carloway, which offers accommodation.

Blackhouse in Arnol with curtained box bed and fire in the middle of the floor.

Those in the district of Back who are of a certain vintage will recall that Mina's parent's blackhouse was the last inhabited one remaining in the district. I often feel privileged to have had the opportunity to visit her mother and father there when I was young. My late mother, who was closer to Mina's age, recalled what life was like for them as children, growing up in Coll and living in a blackhouse. She remembers it as being "very cosy."

We had oil lamps and candles for light, but we didn't have an indoor toilet, bathroom or shower. We had to go outside to do our business, and we had our bath in an enamel tub by the fire. There was no running water either, so all the water had to be fetched from the wells around the village. It was nice to go to the well because you always met other people and it was a very sociable chore. We all slept in the same room, in cosy beds with curtains round them, and our mattresses were made of straw. We had large, heavy blankets over us.

We ate everything that was put in front of us and liked it; there were no second choices. Everything was cooked on an open fire; there were no ovens in the blackhouses. We all had hens, sheep and cattle, and many of the men were fishermen, so we ate lots of fish, eggs, vegetables and potatoes. Red meat was usually only eaten on weekends.

We spent most of our free time outdoors, finding bird's nests and getting to know all the names of the birds and the kind of eggs they laid. We knew all the names of the wildflowers.

Most of us wore the same kind of clothes: tweed skirts made by our mothers and hand-knitted sweaters. We also had hand-knitted woollen stockings with elastic garters at the knee. Our footwear was tackety [hobnailed] boots in the winter, but we would get a pair of brown sandals for the summer. I really think we had a very privileged upbringing—we had the moor, the glens, the machair and the beach so close at hand.

We all went to Back Junior Secondary School at age 5. We only had Gaelic, but from the minute we stepped into school we had to learn all our lessons in English, which made it ten times more difficult for us. We didn't have any workbooks. We used a slate and slate pencil, and we had to have two little tins—one for a wet rag to erase our work and another for a dry one to dry the slates.

Mina's father is well-remembered by the older generation here in Coll and Back as the area postman, who in those days delivered the post between Upper Coll and Gress by foot and by bicycle. Back Post Office and Shop were only a short distance from their home, and during the time of the Second World War, Mina worked there as a shop assistant.

Mina was only a few months old when the First World War was declared. Her childhood and young adult life were lived under the cloud of war and rumours of war. I believe that Mina worked in service in Edinburgh for a time. It was common practice for young women to seek employment in the bigger cities, working as housemaids, cooks and kitchen maids in big houses in places such as Edinburgh, Glasgow, Aberdeen and London. The two world wars and emigration efforts removed a disproportionately large number of the young men from the island.

Mina (left) with neighbours Jessie Campbell and Margaret's Aunt Bell Ann Stewart.

Life on Copper Island was a very isolated existence, but Mina's upbringing on the Isle of Lewis meant that she was uniquely equipped, in many respects, for the off-grid lifestyle that she and Nelson embarked upon in Barkley Sound. She truly lived a life on the edge of two mighty oceans. However, there was one major aspect of life on Copper Island that would have been in sharp contrast to the life she had known when growing up in the village of Coll: the close-knit community life and the fellowship of the many friends and relations by which she was surrounded daily.

6. Love and War

From Quebec, Private Nelson Dunkin and the Canadian Forestry Corps were sent over to Britain, where they were based in Scotland. One time when Nelson was on leave, he decided to travel to the Outer Hebrides off the West Coast, as described by Margaret Stewart in the previous chapter. Like many of the young men from Canada who had nowhere to go when on leave, Nelson sought out any connection he could find. He visited the Isle of Lewis, where a friend's mother lived. Mrs. Isabella Paul had married a man from Canada but she returned to the Isle of Lewis when he passed away.

Of course, while he was there, Nelson had occasion to visit the post office in Stornaway to send letters home. But it seems that, having visited, he found reasons to visit again, and again, and again. Or just one reason, really—the lovely post office lady named Mina MacIver. We don't know whether this was a whirlwind romance or how often Nelson had the opportunity to return to the Isle of Lewis. But we know that Nelson and Mina soon decided to marry.

Permission from the commanding officer would have been required, and the army had an official policy discouraging such marriages. But since they were inevitable, the Canadian military assisted young couples with the arrangements. Nelson had both permission and leave to marry Mina on Christmas Day, 1942, in her Black House home near Stornaway, Isle of Lewis. Nelson wore his uniform, Mina wore a simple dress, and they looked happy. A common wedding gift at the time was ration booklets, which a young couple would need for the most basic amenities.

Possibly Mina's Black House home on the Isle of Lewis where she and Nelson wed. Mina is not in the photo, but her father and sister are on the left.

It is unclear what happened in the months after the wedding. Certainly, Nelson returned to his post while Mina remained in Stornaway. As a member of the Canadian Forestry Corps, it is possible that Nelson was transferred to the continent at some point and saw action. The Forestry Corps was typically sent to the front in preparation for an advance into enemy territory. The artillery would barrage a forest until the opposing forces retreated, and the Forestry Corps was sent in to log the area and make it accessible. Most of the timber was used for corduroy roads that enabled heavy equipment to navigate the wetlands.[1] It must have been an extremely traumatizing experience for these men.

Wherever he was stationed and whatever happened to him, the ordeal had a permanent effect on Nelson. "There was a whole period that he was completely silent about," Nelson II recalls. "I guess they would call it Post Traumatic Stress Disorder. But that was a closed book as far as us children were concerned. He just didn't talk about it." When I mentioned

Nelson and Mina's Wedding
Christmas Day, 1942

this to Margaret, she wasn't surprised. "I hear that so often, people saying, My father wouldn't talk about the war. And I say, Are you surprised? Would you want to talk about that with your children? No, I'm not surprised he wouldn't want to talk about it."

As a result of the trauma of the war, Nelson developed a nervous condition, evidenced by a recurring twitch in his face and a nod of the head that many remember. He told me once that he had no sense of smell—and therefore little sense of taste—due to being gassed in the war. However, neither gas nor chemicals were used in the Second World War, only in the First, and these agents didn't typically destroy one's sense of smell. So what he was referring to is unclear. One of the effects of severe shell shock can be the loss of one's sense of smell, along with nervous disorders.

In any case, the Canadian Military returned Nelson to Canada in the fall of 1943 and stationed him on Vancouver Island. Later, he would be honourably discharged and receive a pension. However, when he inquired about returning to the USA, he was turned down. You enlisted in the Canadian Army, the Americans reminded him. You are a Canadian now and no longer welcome in your home country.

Mina Dunkin, like most war brides, was not permitted to accompany her discharged husband to Canada, and remained in Stornaway until a year later. If Nelson's later love for snail mail is any indication, we can imagine that many letters went back and forth. Finally, the Canadian government made arrangements for Mina to join her husband. The Canadian Veterans website describes what Mina would have experienced:[2]

Many war brides describe receiving just a few days' notice before it was time to sail for Canada. There were often heart-wrenching scenes as young women said goodbye to their families.

War brides were transported on huge troop ships especially outfitted for their use, and converted luxury liners. The most notable of these was the Queen Mary. War brides remember sharing the ship with nearly 1,000 other war brides and their children. Red Cross Escorts did their best to ensure that everyone was taken care of and earned unreserved praise from grateful war brides.

Some war brides describe their voyage to Canada as a great and wonderful adventure. They made friends, feasted on the plentiful supply of food onboard ship, and did what they could to help out those with small children. Others described themselves as homesick, heartsick and seasick.

Almost all war brides vividly remember docking in Halifax, and passing through Pier 21. They were met there by Red Cross and Salvation Army volunteers, who offered the new Canadians a warm welcome and gifts of food.

[They boarded] special war bride trains bound for various points across Canada. Husbands and families were notified of arrival times. Many brides marvelled at the vastness of their new land. For some, the journey to the Prairies and to the West Coast took several days, and seemed almost never ending.

Pier 21 in Halifax.[3]

Mina travelled by Canadian Pacific Railroad from Halifax to the depot in Vancouver (now called Waterfront Station), arriving there on the 13th of October, 1944. Although Nelson was notified of Mina's impending arrival, he was not at the train station to meet her. A short blurb in a news article that day tells the story:

> *Mrs. Nelson Dunkin, wife of Pte. Dunkin, now stationed on Vancouver Island, hoped to see her husband in Vancouver today—almost exactly one year after he left her in England to return to Canada. Unfortunately, Pte. Dunkin couldn't get leave to come to Vancouver.*[4]

Margaret said that Mina had several relatives living in Vancouver who might have met her at the station. "There were a lot of MacIvers from both sides of the family who lived in Vancouver. It gets a bit confusing, and when you're young, you just know they're all relations but quite confused about who you belong to. All my cousins in Vancouver seem to be called MacIver." Likely, it was Aunt Marion and Uncle Calum MacIver who welcomed Mina to the West Coast of Canada.

But try to imagine the end of this long journey for Mina. She has travelled for five days and 5000 km from England to Halifax, and about the same distance and nearly a week by train to Vancouver. She hasn't seen her husband for a year, and he is not able to be there when she arrives. She is taken home by relatives whom she doesn't know. And when Nelson is finally able to join her in Vancouver, she sees that the war has made him a different man than the one she married. She has no option but to see it through, and her ability to do this shows remarkable resilience and faithfulness—and enduring faith in God.

---

[1] Robert Briggs, "Canadian Forestry Corps in WWII," freepages.rootsweb.com/~jmitchell/genealogy/cfc54.html

[2] "Canadian War Brides," veterans.gc.ca/eng/remembrance/history/second-world-war/canadian-war-brides

[3] Canadian Museum of Immigration at Pier 21 Collection (DI2013.1205.1)

[4] "War Brides Greeted by B.C. Kinfolk," *The Vancouver Sun*, October 13, 1944

7. A Tale of Three Ports

Once Nelson had been discharged from the army, he and Mina moved into a basement suite at the home of her aunt Isabella MacIver on Sophia Street in South Vancouver. Nelson began work at the Canadian Fishing Company (or Canfishco) in Steveston, each day taking the electric Interurban Tram Line[1] down Granville Street. Nelson worked with the shipbuilders, and here his ingenuity soon came in handy again. Seiners now use a drum system to set, draw and store the nets, but at that time they used revolving seine tables, which were turned manually. Nelson came up with a system for turning them mechanically. The company began sending him to their other locations to install and maintain his invention. He often took the tram to Burrard Inlet. Or he would catch a ride on one of the fish-packer boats to Port Albion, across from Ucluelet on the West Coast of Vancouver Island.

One day, Nelson was in Port Albion installing the new seine table equipment on one of the boats. His helper accidentally dropped his end, resulting in Nelson breaking his back. For a time, he was not able to work on the boats. Rather than being idle, he began building a float house for himself and Mina at Port Albion. He continued travelling between there and Vancouver as he was able.

During one of those trips home, Mina told Nelson the happy news that she was expecting. However, it was a difficult pregnancy, and Mina was unable to leave Vancouver and go to their new home in Port Albion. On August 5th, 1946, Mina gave birth to twins. Sadly, the first to be born did not survive. "At that time, my mother cried out to the Lord and said, 'If you give me the second one, I will give him to you, to preach and minister the rest of his life," Nelson II explained. "I guess I didn't have a choice." Nelson II, the second twin, is still a preacher of the Gospel to this day.[2]

DUNKIN—Born to Mr. and Mrs. Nelson E. Dunkin (nee Jamesina MacIver), of Port Albion, Vancouver Island, B.C., on August 5, 1946, at Vancouver General Hospital, a boy. Both doing well.

For the first two years, the young Nelson II vacillated between life and death. Mina had to remain in Vancouver to be close to medical help. Nelson continued working in Port Albion, returning to Vancouver when he was able. Finally, Mina and son were ready to move to their new home in Port Albion. One of their neighbours, Lloyd Bridal, remembers Nelson's kindness to him there as a child:

> *Back down, as you turned right along the path, was a floating house where I believe the machinist lived, Nelson Dunkin. And his wife was Mina. I remember telling them about a little steam engine that was made, and you could just lay a little fire in it and get it to operate. And he says, Well, why don't we do that? So he went into his machine shop, and he made up the dandiest, slickest little steam engine. And then he gave me a coffee pot with some tubes. I was supposed to work out a way to get these tubes to fit in the coffee pot so it would be the means of heating the steam for the steam engine. I felt he spent a lot of time and took a lot of interest in us boys.*[3]

At that time, Nelson also crafted a steam-powered tugboat that Nelson II enjoyed for years. It was good to finally settle in together as a family. Soon after she and their son arrived in Port Albion, Mina gave birth to a daughter on August 4th, 1948 in Tofino—the nearest hospital, 40 km away. She was named Madge after Nelson's sister who had passed away so young. "I always look for people who are born in Tofino," Madge told me. "I'm like, 'Whoa! So was I.'"

The family lived in Port Albion for two more years. Then Nelson took a job with BC Packers, and they moved to Port Alberni for a short time. Nelson II remembers that they lived in a float house at Polly's Point on the edge of town, near a lighthouse and next to a First Nations reservation. It is unclear if this float house was the one from Port Albion or a new one. In any case, this move was short-term. The place where Nelson wanted to live was a distance down the Alberni Inlet in the once-booming community of Kildonan. Their relocation to this village would be the start of the happiest time in the life of the Dunkin family.

Madge and Nelson with friends on the wards of the Kildonan Cannery.

---

[1] You can visit one of the trams that Nelson likely rode, on display in Steveston. stevestonheritage.ca/visit/steveston-tram/

[2] Inset: "Births," *The Vancouver Sun*, August 10, 1946

[3] Interview with Phil Hood, 2023

The wild West Coast of Vancouver Island. Can you spot the natural stone arch that Russell took us to see?

Barkley Sound from above Clifton Point, Alberni Inlet northeast in the distance.[1]

[1] Drone photography by Brandon Day

8. Barkley Sound

We are backing up in time to look at the history of Barkley Sound, which was the main stage for the better part of Nelson's story.

The West Coast of Vancouver Island was once populated by the Nootka peoples, now called Nuu-chah-nulth. The Nootka once comprised many tribes; today, the Nuu-chah-nulth tribes are 15 in number. In Barkley Sound, this includes the Yuułuʔiłʔatḥ tribe across from Ucluelet, the Toqhuaht band near Toquart Bay, the Uchucklesaht tribe near Kildonan and the Huu-ay-aht people at Pachena Bay.

These tribes were sustained by the abundance of game, vegetation, shellfish, salmon, halibut, seals and whales in Barkley Sound. They were the only West Coast people who regularly engaged in whaling, which was a significant part of their culture. A single whaling expedition required months of preparation and ritual. The various tribes were generally peaceful, but there were occasional clashes.

On January 26, 1700, a huge earthquake shook the entire West Coast of North America and caused a massive tsunami. The wave devastated the shores of western Vancouver Island. The entire village of Luh'taa near Pachena Bay, among many others, disappeared overnight:

> *While the people in Luh'taa slept, the earth began to shake. The houses on the sandy beach swayed and the cedar boards shifted. By the time the people awoke—if they awoke at all—it was chaos. Beneath them, the sand shifted and turned to liquid; all their houses flattened and fell into the sand. There*

was no time to get into the canoes, and no way to make the distance from their houses to the water, for now all was liquid. Canoes floated out to the ocean while the villagers slid beneath the sand. And soon, very soon, the tide that pulled back from the beach and released the canoes brought into the beach an enormous wave.

The wave, which must have glistened in the moonlight, crested and crashed as it hit the land, its power splintering the mighty fir and spruce trees in the forest well beyond the village site. As it sucked back with a tidal strength never before imagined, the wave took with it the forest, the animals, the rocks and the entire village. All were killed; none remained living among those who had gone to sleep that night in the village of Luh'taa.[1]

But the greatest devastation was yet to come. The Nuu-chah-nulth were among the first indigenous people that European explorers encountered on the West Coast. Within 50 years of first contact and trade with these explorers and settlers in 1774, disease and conflict had decimated 90% of the native population. The first of the explorers were the Spanish—Juan Hernandez and Esteban Martinez—and the British—James Cook and John Meares.

In 1786, Frances Barkley stepped aboard the *Imperial Eagle* with her new husband, Captain Charles Barkley. She was 17 years old. Leaving the River Thames, they would sail for eight years around the Pacific Rim, including the west coast of Vancouver Island, and Frances would be the first European woman to step ashore there.[2] They were seeking sea otter pelts, which were of great value to the Chinese. Charles named the large island-studded Sound after himself and other prominent landmarks after members of his crew.

For a few decades, the Nuu-chah-nulth frequently interacted with European and American traders, acquiring many goods they had never seen before, especially metal. But there were soon few sea otters to be found. There was little further trade until the 1850s, when dogfish oil was in high demand for lubrication in the sawmills. Ben remembers Nelson telling him that dogfish oil is an ideal fine oil for machinery. The Nuu-chah-nulth industriously caught and processed large quantities of dogfish for this purpose. Fur seal hides were also sought by the traders.

European settlement was slower in Barkley Sound than on the East Coast and Southern Vancouver Island. By around 1860, Captain William Spring of Victoria had established trading posts all along the West Coast, trading primarily for dogfish oil and seal furs. In the summer, Andrew Lang managed a post at Dodger's Cove on Diana Island in the Deer Group Islands. In the winter, he managed another trading post at Clifton Point on Copper Island, the property that Nelson would purchase a hundred years later. R. Bruce Scott remembers that Nelson had "a survey map show[ing] the location of the original trading post buildings and wharf."[3]

I have often sat at Pebble Beach or Prayer Cove on Copper Island, running my hands through the clean, fine gravel in the hope of finding glass beads. They are usually blue-green; I don't know if it is because all the trading beads were of that colour or if the blue-green beads were the only ones to survive. They are a connection to 150 years past, a history that is both intriguing and sad. I treasure the few beads I have found, and I imagine the Huu-ay-aht did as well. But those cheap beads also funded the lavish, work-free lifestyle described in stories like *Pride and Prejudice*. My glass beads were among the many little reasons that those British gentlemen and ladies had nothing to do but marry off their daughters to other rich men.

Copper Island was witnessing an increase in traffic. An American brig, the *Swiss Boy*, made an emergency beaching in Robber's Pass at the south end of Copper Island in 1857. "As the *Swiss Boy* and her crew were American (or 'Boston Men'), the boat and its contents were claimed as a foreign intruder's ship by members of the Huu-ay-aht and Tseshaht First Nations who had loyalty to the British ('King George's Men'). Although no one was harmed by this 'plundering,' as these actions were dubbed by British officials, this became known as the '*Swiss Boy* Incident,' and led to the appointment of William Eddy Banfield as Indian Agent of the area to 'keep the peace.'"[4]

Above the door of the trading post on Copper Island was the name board of the 1067-ton ship *Orpheus*.[5] This boat was infamous for one of BC's worst marine disasters. In 1875, the captain of the *Orpheus* had mistaken the *SS Pacific*'s single masthead light for the Cape Flattery Lighthouse, and as a result, they collided. The *SS Pacific* went down, with only two out of an estimated 275 passengers surviving. The *Orpheus* was able to continue its journey, but its propeller was damaged, and it ended up on the rocks on the southwest corner of Copper Island.

There was also growing interest in what lay beneath the ground on Copper Island. "In 1860, Bamfield trader and government agent William Eddy Banfield had reported copper on Copper (Tzartus) Island in Barkley Sound. While still at the Alberni sawmill, and collecting taxes, supervising tugboats and shipping, and a boatyard, Capt. Stamp formed the Barkley Sound Copper Company and removed seven tons of ore from there. Free of his mill job in 1863, Stamp and his sons scoured the Island and also some copper showings on Santa Maria Island in Sarita Bay, but to no avail."[6]

The Church was soon to follow. In 1877, the Reverend A.J. Brabant established a Catholic mission at Namukamis, near the mouth of the Sarita River. His journal reads, "I left Victoria on the schooner '*Favourite*,' Hugh McKay captain, on the 23rd of August, accompanied by a French-Canadian carpenter called Morrin, and arrived the next day in a small bay on Copper Island opposite the Sarita Valley and river. From there we went and carried in canoes our provisions and tools, and selected a spot for the buildings close to the Namukamus Village."[7] The bay that Nelson would daily look out on and affectionately call "Sunrise Cove" made a sheltered base of operations for building the mission. The missionaries sought to counteract the immorality introduced by the fur traders but brought some detrimental influences of their own.

The wealth of timber on Copper Island became the next interest. In the summer of 1907, John Webb Benson logged about one million board feet of timber at the north end of Copper Island. Unfortunately, after doing so, he could not find a buyer. In subsequent years, large parts of Copper Island were clear-cut. Nelson's property remains one of the few areas with ancient old-growth forest, which dwarfs the visitors who wander through it.

---

[1] Kathryn Bridge & Kevin Neary, *Voices of the Elders, Huu-ay-aht Histories and Legends*, 2013

[2] For an intriguing account of this journey, see *The Remarkable World of Frances Barkley: 1769-1845*, by Cathy Converse and Beth Hill, 2008

[3] R. Bruce Scott, "Independent People," *The Daily Colonist*, February 9, 1969

[4] Facebook post, Bamfield Historical Society, January 27, 2017

[5] Jan Peterson, *Journeys Down the Alberni Canal to Barkley Sound*, 1999

[6] "Stamp Strikes Out in New City of Vancouver, *Alberni Valley Times*, May 6, 2008

[7] Rev Chas. Moser, *Reminiscences of the West Coast of Vancouver Island*, 1926

The Dunkin family in Kildonan in 1957.

9. Kildonan

I like to think that Mina was happy to move to Kildonan, nestled in Uchucklesaht Inlet off the north shore of Alberni Inlet. The village was named after a town on another Scottish island south of her own. Also, Kildonan was still a bustling community in 1950; perhaps 10 to 15 families plus other individuals called it home. Though the Cannery closed in 1946 after an earthquake destroyed several buildings, fish were still processed at the reduction plant. However, as Nelson and family arrived, fish stocks were becoming depleted on the West Coast. "I saw the tail of the last pilchard going through the tally machine," Nelson told a reporter years later.[1]

Nelson had purchased a mineral claim around the corner from the post office in Kildonan, in what he named Dunkin Bay, next to what is locally called Huckleberry Point. The property had belonged to an eccentric Remittance Man[2] named Bains, who after the war was fearful of another invasion. "The guy had made trenches and fortifications, and used to march around with a shovel over his shoulder," Peter told me. Because of the nature of this mineral claim, Nelson owned even the property under the bay, and for perpetuity.

If you take the M.V. *Francis Barkley* from Port Alberni to Bamfield, you will likely stop in at Kildonan. The ferry turns into Uchucklesaht Inlet, circumvents a little island and pulls up at a small float that houses the post office. On the way, you pass hundreds of pilings, both on shore and in the water. This is where the Cannery once stood, as there was no flat land to use for its construction. In 1962, all production ceased there, and the Cannery was torn down and burned. Only the wharf and the house of the manager were spared. When Nelson was told of the impending demolition, he rushed over at the last minute and salvaged all the ropes, blocks and hardware he could find, tossing them into his boat.

Nelson had his float house towed from Polly's Point to the new property in Kildonan, and that is where the family lived. Nelson's boat, *Raven*, was tied up alongside. On shore, the family kept a few goats and chickens and kept a small garden. Madge remembers these as happy years for the family. There were about 20 other children in the community, a one-room school to attend and events to enjoy. "Mina would have been used to living in a community where everybody knew one another and was in and out of each other's houses," Margaret told me. "We're still the same here [on the Isle of Lewis]. We know everybody around us in all the villages in this district." Kildonan was a great place for the kids to grow up.

The Dunkins' float house was very small. A *Life Magazine* article recounts a time when the Shantymen's boat, *Messenger III*, visited the family there at Christmas. Earl, one of the missionaries on board, told me, "I knew Madge when she was four and young Nelson was six. Precious young people, like everyone… I think God gave a piece of himself to all of us when he breathed into us the breath of life." The article describes the Dunkin home as "tiny" and even too small for the Christmas tree the missionaries had brought for the children.[3] The house was sheathed in tar paper, and Nelson's name for it was "Old Ugly."

Perhaps for this reason, Nelson began constructing a large cedar building on shore, variously described by those who saw it as a warehouse, a multi-storey home and a cabin. Peter's description of it as a cabin is probably the most accurate, as he worked on it and the building is still in his community. "The way he built it was amazing," he told me. "With split logs and total Nelson style. As on all his cabins, he used a method of putting on shakes that is very efficient and distinctive. Instead of a multilayer, it's a single layer. I've used his method ever since, and it takes way fewer shakes and seems to last forever." But the family never lived in this building. When they later left Kildonan, Nelson generously invited his friend Elmer Matthews to take it over.

BC Packers moved Nelson around to do repairs on the seine boats for a short time. But soon all fish processing ended, and the only remaining operation at the Cannery was to supply ice for fishermen. Nelson took on the job of watchman for the Cannery and became a Fisheries Patrol Officer. He also continued to innovate: Nelson designed a new hook to connect the chains on log booms. McMillan Bloedel used his invention in their boom works. Everyone asked Nelson, Why didn't you patent that? But it wasn't Nelson's way.

Peter of Kildonan remembers Nelson as the first in the community to attempt an Alaskan mill, a system that uses a chainsaw to mill lumber. "He was the grandfather of Alaskan mills in Kildonan," said Peter. "Everybody here had the Alaska mill at one point or another. But Nelson is the one who figured it out and showed us how to do it. We were just kids, and we were so interested in the tons of stuff he could teach us."

Nelson was a leader in the community. During the 1962 Federal Election, Nelson served as Deputy Returning Officer. Ballot boxes for isolated areas on the West Coast were delivered by the Federal Fisheries Department boat, the *Comox Post*. Nelson was responsible to receive the ballot box in Kildonan, set up the polling station, count and report the votes, and later return the ballot box to the boat for pick-up.

"Ballot boxes for isolated areas of Vancouver Island are delivered by federal fisheries department boat, *Comox Post*. Here Deputy Returning Officer Nelson Dunkin of Kildonan, Barkley Sound, takes his box ashore for Monday's election."[4]

Peter remembers, "One day Nelson went out, and here was one of his goats—which had horns, right? And there was a young girl, I guess from a family that was visiting or whatnot, and the goat had this girl pinned up against the wall with its horns on either side of the girl's waist. Nelson just lost it. He grabbed the goat, grabbed the hacksaw, dragged the goat down to the beach and sawed both horns off.

"Everything moves on Kildonan time down here, right?" laughed Peter. But as much as the Dunkin family was happy with Kildonan life, Nelson was restless for more. He discovered that one of Captain William Spring's ancient trading posts was up for sale on nearby Tzartus Island, locally known as Copper Island. This was R. Bruce Scott's description: "The property contains 105 acres, mostly waterfrontage… Geographical features include a lovely sandy beach with a waterfall at one end, and the forested slope of Copper Mountain containing several mineral claims, some still in good standing."[5]

Nelson's mind raced with the possibilities. What would be his purpose in securing this isolated property? From conversations with him, I believe he always had in mind to build a community there. Brian remembers that a trade of some sort was involved; R. Bruce Scott records that the land was part of an estate sale. However it happened, Nelson became the new owner of the property surrounding Clifton Point on Copper Island. At the same time, he bought little Davies Island to the north, which he later resold.

Nelson immediately began preparing timbers for a home on the island, loading them onto a log float. He often boated over to his new property to move drift logs and prep the site for their float house and for the new construction that would follow. This would be a turning point for the Dunkin family. For Nelson, there was no turning back.

Waiting for the Lady Rose with neighbours on the dock in Kildonan.

"Nelson pushing a cart on the wards of the Kildonan Cannery.[6]
My neighbour and I found that very cart, or one similar to it,
when we were cleaning up the grounds." - Peter

---

[1] Katie Poole, "An Island of Solitude," *Tri-City News*, 17-Nov-1991

[2] The black sheep of British families, sent to the Colonies to live on an allowance.

[3] "The Mission of the Shantymen," *Life Magazine*, January 11, 1954

[4] Front Page, *The Vancouver Sun*, June 15, 1962

[5] R. Bruce Scott, *People of the Southwest Coast of Vancouver Island*, 1974

[6] Photo (cropped): Alberni Valley Museum Photograph Collection, PN2580

10. A Christmas Adventure[1]

by R. Bruce Scott

The following is an account written by R. Bruce Scott in 1974 after visiting and interviewing Nelson and Mina Dunkin at their home on Copper Island. The story is very well told, and I was at a loss as to how to receive permission to include it in full here. I knew that his daughter lived in Victoria, but the address in the book was 50 years old. I thought, hmmm... Oak Bay address, older family home... what are the chances that his daughter still lives there? I sent off a letter and a week later received a call from Susan M. Scott. She kindly and enthusiastically permitted me to reprint her father's amazing story, which occurred shortly after the arrival of the Dunkin family in Kildonan.

INK DRAWING BY SUSAN M. SCOTT

It would be difficult to imagine a series of adventures more hazardous or bizarre than those of Nelson and Mina Dunkin, and their two children, aged two, and four, while returning from a Christmas shopping expedition down the Alberni Inlet in December 1950.

Only recently having established themselves in Kildonan, a small village twenty miles down the inlet, they were, as yet, comparatively inexperienced water-wise, and unfamiliar with the dangers of navigation—dangers that lurk even in the comparatively sheltered waters of the inlet.

So it was that just a few days before Christmas they blithely set out in their small gasboat, which Nelson used for commercial fishing in the inside waters of Barkley Sound, to do some Christmas shopping in Port Alberni, at the head of the inlet, taking their two small children with them.

After they had finished their shopping, they left Port Alberni late in the afternoon and headed down the inlet for Kildonan. By the time they reached Nahmint, halfway down the inlet, it was dark and blustery, so Nelson decided to anchor in the mouth of the river for the night. Early the next morning, although it was still raining and blowing, they started out again. They had not gone far when the boat hit a deadhead, a water-logged log that floats perpendicularly with one end invisible just below the surface, the other resting on the bottom if the water is shallow, floating if it is deep. In this case the log was resting on the bottom and, try as he would, Nelson could not dislodge the boat, which was holed and stuck fast. Fortunately they were towing a skiff. They also carried lifejackets. Fastening the lifejackets on the children, Mina climbed into the skiff. Nelson handed the children to her, then threw a bundle of blankets into the skiff. These and a small black bag containing a Christmas cake which a friend had given them as a parting gift, and a loaf of fresh home-made bread, were all that they took with them.

Mina waited while Nelson searched the boat for something which he did not find, then, seeing that the boat was about to sink, called to him, "You'd better jump off." With no time to lose, Nelson took an axe, cut the rope that tied the skiff to the boat, and clambered in just as the gasboat sank beneath him.

Not knowing what to do, or where to go, they landed on the nearby shore. It was pouring with rain and the wind was still blustery. Remembering that there were some deserted Indian shacks farther up the river, Nelson decided to row there in search of shelter. By this time they, and the blankets, were soaking wet. Choosing the least dilapidated looking of the shacks, he lit a fire on the earth floor and tried to dry the blankets. It was obvious that they could go no farther in their twelve-foot skiff until the weather improved, so they prepared to spend the approaching night where they were.

Intending to serve a piece of Christmas cake to each before they turned in for the night, Mina opened the black bag and found that both the cake and the bread were soaked with saltwater and quite inedible.

Nelson found an empty fishbox (a strongly built box measuring three feet by two feet, used for packing fish for transportation to market) in the shack, so, turning it open side up, he lined it with blankets and used it as a cot for the children.

Sleep was out of the question. They tried to rest, but the roof leaked and the shack was filled with smoke from the fire, which they had to keep burning for warmth. The children cried restlessly. Nelson and Mina spent most of the night standing in the doorway to clear their eyes of smoke. They couldn't go outside because of the pouring rain.

When daylight came they were a sorry sight. Their faces were blackened by smoke, and their eyes red and sore. As Mina said, they looked just like smoked salmon.

"There's another shack farther up the river on the other side," said Nelson. "The Boy Scouts are building it for their summer camp. Let's go and try that one, it should be in better condition than this one.

They all clambered into the skiff and Nelson rowed upstream to the site of the Scout's camp. On arrival, they found that it was still unfinished and only partly roofed. It had a shed roof and there were two rough bunks under the eaves. Disappointed at what they saw, they had no option but to stay the night and hope for the best. But it wasn't any better than the previous place. They lit a fire and it smoked copiously, billows of smoke filling the shack unbearably. They had to contend not only with smoke but, as before, with a leaky roof; rain dripped through the roof and soaked them all thoroughly. They spent another sleepless night and, to add to their worries, the rain-swollen river flooded its banks and inundated the valley. All night long they heard the roar of rocks, loosened by the heavy rain, which thundered down the cliffs of the ravine—never knowing when one might roll on them. "It was just like thunder," Mina said, "just like thunder all night long."

When daylight came, Nelson saw an old Indian dugout canoe that had been partly chopped up, apparently for kindling. Both sides had been chopped half way down. Nelson figured that it could be made serviceable by stuffing the cracks with rags, but where to find rags in that isolated place? His wife's dress, of course! They used the canoe as a dingy and stowed what few possessions they had in it in order to lighten the load in the skiff.

At long last the storm abated and the day became really calm. Knowing that there was a logging camp farther down the inlet at Long Bay, Nelson decided to head for it and stay there for the next night. Once more they set out. It was hard work rowing a skiff-load of humanity and towing a heavy dugout canoe, and it took a long time to cover even a short distance. As they emerged from the mouth of the Nahmint River into the Alberni Inlet, they saw a fishing boat coming down the inlet from Port Alberni. Nelson tied his white shirt to an oar and, standing up in the skiff, waved it vigorously to attract the attention of the boat owner. "He flagged that boat, and he flagged it, and flagged it," said Mina, "but it never changed course. We'll never know whether he saw us or not. Surely there had to be a reason! Anyway he went right on by."

Finally, after long hours of rowing, sometimes against the tide in the eddies that whirled along the shoreline, sometimes with it, sometimes with the wind, sometimes against it, they eventually arrived at Grumbach's logging camp at Long Bay. The place was in darkness and there was no sign of life when they knocked at the door. There was no response so, leaving Mina and the children at the front door, Nelson went around to the back to see if there was any unlocked door or window that they could enter and take shelter for the night. While he was away the door suddenly opened and a tall man stood in the doorway and asked sullenly, "Well, what do you want? What are you doing here?"

Hearing the sound of voices, Nelson returned and told the man what had happened and that they would like shelter for the night. At first he refused, then seeing their bedraggled condition, reluctantly let them in. Cold, wet and hungry, not having eaten since their boat sank, Mina asked if they could get something to eat. At this he hummed and hawed. In desperation, Mina assured him that they knew the Grumbacks, owners of the camp, very well and felt sure that they wouldn't mind if she had a look around to see what she could find in the way of food. So he relented and told them to go ahead.

While the Dunkins rummaged for food, the man apparently the caretaker seated himself in an armchair with two cases of beer and two jugs of wine within arm's reach. "He had a wonderful time," said Mina. "He was a Swede, at least six feet tall. Obviously he had been drinking before we came, so he carried on where he had left off."

She did not say what they found to eat but they did find something and, while they were drinking their tea at the end of the meal, the man started swearing at them. He said he was a communist and launched into a long diatribe about what they were going to do to the capitalists in this country when the opportunity arose. The Dunkins suffered his outburst in silence then, in order to quiet him, said it was time for them to go to bed. But the man would not let them go. He continued the harangue. Finally, around ten o'clock, he told them they could go over to the bunkhouse to sleep.

"But do you know what he did?" asked Mina. "He told us to get up, then he lined us up in order with the smallest child first and Nelson at the end. He stuck a rifle in Nelson's back and said threateningly 'Go on! March down to the bunkhouse!' Can you imagine? The stupid man! But we had to do what he said, so we walked over to the bunkhouse and when we got there he was reluctant to let us go. He didn't want Nelson to stay with me and the children, so he lined us up again and marched us back to the camp. There I laid blankets on the kitchen floor in front of the stove and we all laid down to rest. At two o'clock in the morning I heard a rustling noise. Looking up I saw this man in the doorway. He was looking into the kitchen where we were sleeping, and then he picked up his rifle off the floor. I jumped up and asked him what he was doing. 'Oh,' he said, 'I'm just going to look for the Grumbach's dog: he might bite you.' With that he disappeared but, of course, we didn't sleep much that night."

As soon as daylight came the Dunkins arose, determined to get out of there as soon as possible, regardless of what might happen. Loading their few possessions into the dilapidated dugout canoe, they climbed into the skiff and cast off. By the time they reached Hell's Gate, a notoriously rough stretch of water where the wind funnels between steep cliffs, the wind had risen and there were whitecaps. Desperately Nelson headed for shore, but before he could make it the boat swamped and sank beneath them, leaving them all bobbing on the whitecapped waters.

"I don't know how long we were in the water," Mina related afterwards, "I had been sitting in the stern of the skiff with Madge on my knee. When the boat sank, I lost sight of Nelson and the boy. Fortunately the children had their lifejackets on. Presently Nelson swam over to me and said 'Grab the canoe!' So I placed one arm over the gunwale and held on. Nelson lifted the boy into the canoe first, then he got Madge up and then he struggled to get me in, but he couldn't. I was too heavy. He got in himself and grabbed me and tried to haul me in that way. I was part-way in when a wave washed me out again. Nelson caught me by the leg and held on until he could pull me in again. And, do you know? You'll never believe what happened! The canoe had been tied to the rowboat, but when the rowboat sank, Nelson had to let go the oars to grab the boy. The oars drifted away and we thought they had been lost, but do you know? When we got into the canoe, there were the oars in the canoe, just as if someone had placed them there! It was a miracle of God because it wasn't possible any other way!"

"Yes," Nelson added, "there was about six inches of water in the canoe, and there were the two oars floating inside the canoe." He shook his head from side to side in incredulity.

After bailing out the canoe, Nelson rowed behind Limestone Island for shelter. There, to their joy, they saw the familiar figure of Arthur Maynard, troll-fishing in the bay. Nelson rowed over to him and related briefly what had happened. Maynard invited them aboard and, after making them a hot cup of tea, took them home to Kildonan.

"That was our Christmas shopping expedition," said Mina. "I couldn't swim, neither could Nelson — not more than a few strokes. We can't but see God's marvellous help in that experience."

"Amen," added Nelson.

- R. Bruce Scott

Nelson II and Madge told me about their own memories of this adventure:

I asked Madge, "Do you remember the boat going down at Christmas?"

She replied, "No, thank goodness. I know I'd probably be really traumatized. The capsizing, that's too long ago" (she was 2½ years old).

Nelson II, who was 4½, remembers it clearly. "We had a whole load of our Christmas shopping on board. We were towing the skiff behind us as a safety thing. We were heading back and hit a deadhead at Nahmint Bay and the boat went down. If I remember, it went down pretty quick. I guess they decided they were going to row the skiff from there to Kildonan. I think there was a dugout canoe at Nahmint and he loaded it up and took it. And he towed the canoe—with us in it—behind the skiff. I remember my sister and I sitting in the canoe as little kids. It was leaking, and we were sitting in the water. It was a nice sunny day, believe it or not. And we had a box of mandarin oranges, and that's what we were eating as we sat in the canoe."

[1] From R. Bruce Scott, *People of the Southwest Coast of Vancouver Island,* 1974. Reprinted by permission of Susan M. Scott of Victoria, BC, by phone on July 4, 2023

11. The Move to Copper Island

The official name of Copper Island is Tzartus, which is a Nootka word for "the place of the seasonal or intermittent waterfall."[1] Ben remembers, "One of my favourite places on Nelson's property was the trail down to Pebble Beach. It spoke peace to me. Huge overhanging trees. Great camping spots; we had a youth camp there one time. There was a grotto at the far end of the beach where the creek formed a waterfall. Nelson once reported seeing a sea lion in there."

The author's wife and son at Pebble Beach, 1987

The grotto has since tumbled down, burying the waterfall under huge boulders. But the creek is still seasonal. That's one of the strange features of these coastal islands. Even though nearby Henderson Lake is considered the wettest spot in North America, receiving 7.3 metres of rain per year, the quickly-depleted water supply in the summer is always a concern to island dwellers in Barley Sound.

And the Dunkins were about to become island dwellers. Nelson II remembers that before the move, the family often took the *Raven* to the property on Copper Island to do preparation work. "He cleared some land right on the edge, and he put in logs to skid the parts of the house up to where he wanted it, and made everything ready." Earl recalls that Nelson set bumper logs in place as a buffer between his floats and the beach. "The logs were gathered to keep the floats from hanging up on the rocks. The beach was pretty steep right there."

In early spring 1958, the Dunkin family was ready to make the move to Copper Island. No boxes or moving vans were required—it would be a simple case of towing their float home from Dunkin Bay in Kildonan to the new location, about twelve kilometres distant. To accomplish this, Nelson called on the help of his good friend Earl Johnson, a Shantymen missionary and one of the skippers of *Messenger III*.[2] I will let Earl, who is now 94 years old, tell the story himself in his engaging style:

"As I recall, there were two floats and three buildings: the little house they lived in, and a float with two shacks for tools and equipment and so on. I towed him in the middle of the night because of the freighters going up the Alberni Inlet. Their wash was pretty high, and I was concerned about what it might do to the floats. So we left Kildonan at probably 1:00 in the morning. And if I recall rightly, it was about a two-and-a-half-hour tow from Kildonan to the cove there at Clifton Point. I know that Point well—almost hit it once.

"It seems to me we had the wash of a ship, because when I turned *Messenger* into the wash, I had forgotten about my coffee cup on the bridge, and it fell over. But the rafts survived. Of course, you know Nelson would have had them well-strapped. Everything went fine. Nelson had kerosene lanterns on the floats. And I put the searchlight on the freighter and swung the searchlight around to show the rafts to the freighter, to get them to ease a bit. They were always careful."

"Do you think they slowed down for you?" I asked him.

"Oh, yes. But even then, it was… heh heh… lucky."

The family arrived at Copper Island. Nelson would boat over and continue working in Kildonan for a time, though he was becoming disillusioned by the duplicity of his fellow Fisheries officers. He and Nelson II worked continuously on the new house. "We lived in the float house for a while before the bigger house was built," Nelson II told me. He remembers setting the pilings in place, skidding up the timbers his dad brought with him and building the frame of the house over the pilings.

"And one day, as he was building the house, a big storm came up. I can still remember it," Nelson II continued. The storm came during a very high spring tide. The waves and the high water jerked the float house off its moorings, and the float house began to drift out of the bay—with Madge still on board. "We had the canoe and rowboat out, trying to get lines fastened on the float house. It caught on the edge of Clifton Point, and that was the only thing that kept Madge from going out to sea." As the float house hung on the rocks at the point, Nelson, Mina and their son ran out to get Madge off, along with the dog and the cats.

The float house didn't survive long on the rocks at Clifton Point. The wreckage eventually drifted back into the bay, and the family burned it on the beach. In the meantime, having no other option, the family moved into the unfinished new house. It was just a frame and only partially roofed in. Nelson cut an oil barrel in half and made it into a fireplace, and they made do with an open fire in the house. "We lived in the unfinished house for a long time—nothing was ever a rush with my father," Nelson II recalls. "But I spent hours and hours, and months and months, helping build it."

These rough beginnings were a very traumatic experience for Madge, who was about 10 years old at the time. From that moment on, the last place she wanted to be was on Copper Island. Nelson II, at age 12, was his father's right-hand man and struggled to keep up with his work ethic. "I worked like a dog, night and day," he remembers. And Mina? Only Mina knew how she coped with the hardships the family faced every day. She was known to speak her mind on occasion. However, Nelson II said that when it came to his father, "my mother completely submitted to doing what he wanted to do. But within, she was rather bitter about it."

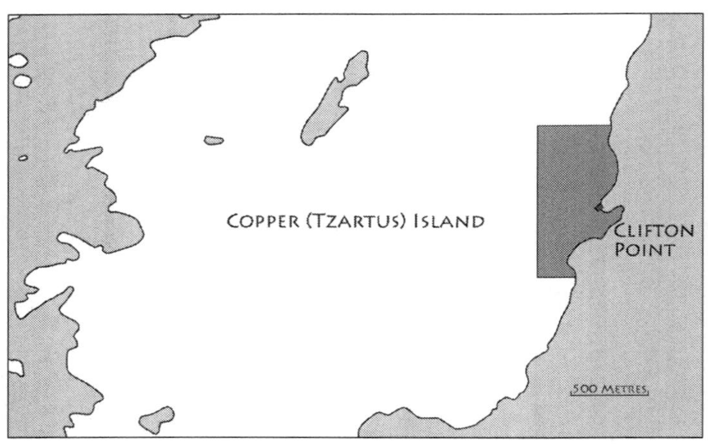

Approximate boundary lines of Nelson and Mina's 105-acre Copper Island property.

---

1 "Tzartus Island," wikimapia.org/18264723/Tzartus-Island

2 Read more about Earl Johnson, the Shantymen and *Messenger III* in a later chapter

12. The Big House

When I first visited Nelson in 1978, the big cedar house he had built was as complete as it would ever be. Two storeys tall plus a dormered attic, with several attachments, the house was built for hospitality. The clearest evidence of this purpose was the dining room table that spanned the length of the main living area. The table could easily seat twenty or more on wooden benches on each side. Plus, there were bedrooms for many people, all with beds and bunks. Nelson built the house to welcome guests.

R. Bruce Scott writes, "For a man with a heart condition, Nelson Dunkin would put many men to shame. After buying the old trading post from the executors of the estate, he began building himself a large house on the shore of the sheltered bay, mostly out of materials gathered on or near the spot. The house was built over the water, so that at high tide the water lapped under the floor and one had to cross a small bridge to reach the shore.

"I asked why he didn't build on shore. Nelson pointed out that it would have meant clearing the land and would have taken too long. The project had started with temporary living quarters on the beach so that they could move in as soon as possible and, like many temporary situations, it became permanent and was added to from time to time."[1]

Nelson II remembers boating all over Barkley Sound, "rescuing" the materials needed to continue building. "We went all around and demolished old buildings and took the lumber. I did lots of pulling nails, which we reconditioned and used in the new house." Peter remembers Nelson as a real packrat: "He salvaged everything." Nelson had also made many friends in the area, and often one or more would come and bring materials, mill wood or give them a helping hand.

Father and son began with a small cabin, which later became the pantry. Next came the main house, with a lower floor that was a single room with the long table. The second floor contained Nelson and Mina's bedroom at the front, plus four more bedrooms (one very small) and a bathroom along the side and back of the house. The attic was a self-contained suite with dormer windows and access to a tiny, fourth-storey lookout beside the chimney. After the main construction came the attached workshop, including a second-storey dormitory above it that could sleep 15-20 people. Finally, another self-contained suite was attached to the forest side of the house. The construction of the Dunkin home was an amazing accomplishment and soon became an iconic feature in Barkley Sound. "It was very big, and it took a very long time," Madge recalls.

The original house, as complete as it ever would be, about 1980.

The Dunkins' sudden need for household furniture couldn't have come at a better moment. Bamfield had been the North American terminus of the Transpacific Telegraph Cable since 1901. In 1959, the cable was extended to Port Alberni, and the Bamfield Cable Station was decommissioned. "The station's luxurious contents and furniture were auctioned and scattered around the region," Darrell Ohs recounts. R. Bruce Scott was responsible for disposing of the station's assets and remembers, "[Nelson] and his wife haunted the station for several weeks, picking up many antique pieces as well as modern furnishings for a song."[2] Darrell continues, "In 1981, I listened to logger and Renaissance man, the late Nelson Dunkin at his home on Tzartus Island, play the ancient pump organ that came from the staff music room."[3] Other furniture came from the closed schoolroom in Kildonan. Nelson II remembers a solid oak kitchen table with sideboard that was used when the family had a few guests.

Visitors recall several unique features of the house. Everyone remembers the flush toilet in the upstairs bathroom.

A rare view of the "back" (front entrance) of the original house, about 1980.[4]

Of course, it flushed directly onto the rocks or bay below, depending on the tide. It had no water running to it, and you were expected to go down and refill the bucket with seawater for the next user. Heather recalls that nighttime flushes were interesting, as the seawater contained bioluminescent algae that lit up the toilet as you poured the water into the bowl.

Rich remembers, "The house was unpainted outside and mostly unpainted inside. You went up this staircase that was right out of *The Lord of the Rings*. On one of the crossbeams was a painting of a duck, and I smacked my head on the beam because it was so low. Nelson pointed to the duck and said, 'Didn't you see the sign?' 'Yes,' I said, 'but it didn't compute until now.' If you didn't hit your head on the crossbeams, you came to the fourth-storey lookout. He said, 'I don't really know why I built it. But isn't it a nice view from up here?'"

The kitchen had running water, drawn from a spring on the hill above the house. The spring still runs today and is the only water on the property that isn't stained brown from running through the cedar woods, such as the water in the creek. The family most often ate at the small kitchen table. There was a pantry directly off the kitchen, a necessary feature since trips to buy groceries were infrequent. The family relied heavily on food they had canned or bought in tins. Mina had a wood cookstove that required firewood chopped small, and this stove shared the chimney with the large woodstove.

Nelson II and Madge don't remember a trap door in the floor of the kitchen, which may have been a later addition. Visitors remember variously the purpose of this trap door. Some recall that it was for plate scrapings and kitchen scraps; others remember that Nelson had built a framework on the rocks below and would drop his tin cans into it through the trap door. Instead of floating away and making a mess, the sea would efficiently rust them away. "We thought it was cool," Bill Irving laughed.

Nelson and Mina had a library, well-stocked with books from everywhere. But what any visitor would remember were Nelson's carvings throughout the house. Every nook and cranny had a plaque or angel or small figures of people and houses. There was the glass cabinet with an upper section depicting heaven and a lower section depicting hell. (More on his carvings in a later chapter.)

In addition to the house, Nelson built several cabins for guests on the shore. Heather remembers Antler Cabin with its natural bucket-flush toilet, and Tom Sawyer's cabin nearby. Still standing today is the Dew Drop Inn near the beach, with barely room for a bed but a great view out the window.

My wife and I remember staying there shortly after the original house burned down. Yes, it burned down. The big house was fabulous but constructed of what was essentially cedar kindling. The loss of the house some 20 years after it was built was a sad occasion, but perhaps inevitable.

R. Bruce Scott aptly summarizes the conundrum of having such a house: "Shortly after Nelson finished building (if a house can ever be said to be finished) and furnishing his home, their son and daughter both got married and moved away, leaving them alone on the island." Even before this happened, one or both of the children were often boarding out in Bamfield or Port Alberni to go to school. How Nelson and Mina would use their home shows that they always considered the place to be much more than their family residence. But before we talk about that, we will take a look at the brief time the Dunkin family lived together on Copper Island.

---

[1] R. Bruce Scott, *People of the Southwest Coast of Vancouver Island,* 1974

[2] R. Bruce Scott, "Independent People," *The Daily Colonist*, February 9, 1989

[3] Darrell Ohs, "They ran a cable round the world," *Times Colonist*, June 24, 2021

[4] Bamfield Community Museum and Archives

Dew Drop Inn under construction, Christmas 1983.
Pencil drawing by Rick Charles, rickcharles81@gmail.com.

The Dew Drop Inn had room for not much more than the bed,
but afforded a magnificent view of the bay.

13. Family Life on the Island

For several years, it was often just the four of them. Nelson and son worked endlessly on the house. Mina and daughter utilized whatever they had on hand to make meals and do household chores. Madge said, "My mother was quite the cook. She could make whatever, something out of nothing. And we always had porridge for breakfast. Oh, yes. How I hated porridge. And if there was any left over, it went into the soup. Well, we were very poor. Put it this way: nothing went to waste." The family kept a garden, with fishing nets strung up to protect the vegetables from the deer, and a few chickens.

Both Nelson II and Madge did school by correspondence at first, with Mina supervising. The children did not have it easy. When I asked Madge about her memories of those days, she replied, "Don't ask me any questions about that. I definitely remember; that's not a problem. It's just that, to me, Copper Island wasn't a happy time. When we lived in Kildonan, that was okay. But Copper Island was not my happy time." The continuing isolation and loneliness were traumatic for Madge throughout the more than five years she lived there.

Nelson II found his father to be a workaholic. "He was always busy, and he had an idea that we would live on Copper Island all our lives, and he had this law laid out for my sister and me. It came to a point where—I guess I was about 13 years old—I was tired of working day and night, night and day. My mother taught me school, and my father worked me. And that was about it. I guess I was always thinking, how can I escape? One day, I was working with him under the house putting in big posts, and one of the posts dropped and I grabbed it or something, and it threw my back out, and I still have the back injury today. I lost it and told him, I quit, I'm set, I'm leaving.

"My mother begged my father to let me go. I said I wanted to go to school in Port Alberni. So they made the arrangements, and I was boarded out in town. Grade eight, Alberni District Secondary School. I lived in three or four different places, at one time with Scottish people who were friends of my parents, and then with Earl Johnson and his family. But my father couldn't understand why anybody wouldn't joyfully live on Copper Island and devote themselves to that for the rest of their lives." In contrast, what Nelson II enjoyed was going to school, joining a peewee softball team and performing in school plays ("Uncle Caris, the head of the family, was played very well by Nelson Dunkin [II]").[1]

The Dunkin family at the small kitchen table of the big house.

Now there were three on the island, except for vacations and some weekends when their son would come home. At that time, Nelson took action on his dream of having a camp on the island, inviting Earl Johnson and the Shantymen to run retreats there. Nelson II remembers going back to the island for some of these retreats and says they were really successful. "For us, though, usually it was work, work, work. Those weren't particularly pleasant times for either my sister or myself. I can understand; I guess my father became obsessive with what was going on, with what he was doing, and we weren't all on the same page."

Madge told me, "Their home was always open, and I mean, they took in people who… well, it didn't matter who you were or what you were. My parents were very generous in that way. But my mother was a people person and would have been much, much happier not living on Copper Island." Nelson II agrees: "She wanted definitely to move to Victoria or somewhere in that area. She wanted to live in a city. And I think she held that against my dad for many years.

"There was absolutely no moving him, whatsoever. My dad had a very definite idea that he was going to create this community where everybody lived in harmony and did everything exactly the way he thought it should be done. And mom was partly in agreement to have the camp; she definitely was. She was a true people person, and he was not. He loved people, but he was a hermit by nature. Mom tried desperately to move. There was a time when they came very close to going their separate ways. But there was absolutely no compromise with him. He just went ahead and did what he did, and you could adjust to it."

Joan of Coastal Missions saw this too. "Nelson had a mind of his own. Now, the Lord used him, and I don't want to be negative, but the reality was, it was Nelson's idea or no idea. And part of it was his escape from the war, the memories of Europe and so on. And so you go hide out. But how would you have reacted if you were Nelson II or Madge? And it's not even with modern conveniences—we're talking about what it was like to live there 60 years ago."

Madge continued doing school by correspondence, with her mother teaching her. For a short time, she went to school in Bamfield and stayed with friends there. Finally, at age 15, Madge had her opportunity to go to school in Port Alberni like her brother, though they went to different schools. The first year, Madge stayed with Earl and Louise Johnson, and at various places after that. She and her brother went back and forth to Copper Island, but it was never home again.

"I was in grade ten, and then I was in grade ten," Madge told me. "And then I got married, at 18." The young man was Andy Vallee, whose home was one of the places where Nelson II had boarded. Earl officiated the wedding. Madge and Andy were married for 25 years and had a daughter, Leona, and two sons, Brian and Matthew, before they were eventually divorced. Madge still lives in Port Alberni, where my wife and I interviewed her. Her kids loved going down to Copper Island and visiting their grandparents.

Their granddaughter Leona remembers those visits fondly. "We would always take the *Lady Rose* down, and I always found it quite interesting because everyone on the boat would pull us aside and ask us why we were going there, how we knew him and lots of other questions. And he would be on the wharf waiting for us with Snuggles. Sometimes, when I was very little, we would stay overnight. I have a few memories of staying in the first house.

"My grandma died when I was quite young. I remember going there, and we would sleep in the upstairs bedroom. She held me and lifted me up onto her high bed, and I would play on it, and I remember her being very loving. They had the wood stove that they would cook on and use for heat. A lot of times, we would stay at the house on the point. Or there were the bunkbed houses just behind where the canoe was being made. We had to get our water from the creek up above the house. There was no power, and I will always remember the kerosene lamps.

"My grandfather would kind of do his own thing, even when we were there. I see some of the pictures with our friends who came along, and we were left to our own devices. It wasn't like he was supervising us. We would walk the trail to Pebble Beach and spend quite a bit of time there, looking under the rocks and stuff. We would fish off the dock and walk all the trails. Sometimes I'd watch him when he was working on his canoe. He had a paddle boat, and we would spend a lot of time on the water in that. Sometimes we'd swim."

Madge said she rarely ever went back for a visit, and her parents rarely came to town. "My father was a recluse in the way he stayed on the island, but he expected everyone to come to him. He wouldn't come out for Christmas, and my mom would be only too happy to be with her family." Madge and Nelson II agreed that, though their parents were different in many ways, they got along very well together and had a close relationship.

For a while, Nelson II was estranged from his father, who had a hard time coping with his family's scattered situation. "He couldn't deal with it very well." But as the years passed, peace and love were restored to their relationship. "I went back to Copper Island," Nelson II recalls, "and we rebuilt the *Raven*, him and I, and made it far more seaworthy. I left high school in my last year, and out of the blue, decided I was going to become a fisherman. I actually lived on the *Raven* and fished out of Bamfield for a time. I was back and forth between Copper Island and Bamfield."

At the age of 20, Nelson II married and had two children, Nelson III and Arlene. Sadly, Arlene passed away suddenly at the age of 12. Nelson II was divorced, remarried, had another daughter—Destiny—and gained several stepchildren. That marriage also ended. He then married Sue-Ann and gained another three stepdaughters. His father thought highly of Sue-Anne, and she and Nelson II remain happily married today.

Nelson II has been actively involved in Christian ministry all of his adult life to this day and was ordained by the Full Gospel Fellowship of Canada. Most of his ministry has been with First Nations peoples, having travelled to every First Nations community on Vancouver Island and many beyond. He has also served overseas in India, the Philippines and other Asian countries, as well as Mexico. He told me, "If you were going to sum it up in one word, my ministry has been one of encouragement. It has covered evangelism, prayer and teaching, but mostly encouragement of people who work in established churches or remote areas."

R. Bruce Scott, a frequent visitor from Bamfield, wrote in 1969, "[Nelson's] wife Mina, who likes the life, admits that it sometimes gets lonely now that the children have gone. Asked what he was going to do with his three-storey house now that they were alone, Nelson said that he was toying with the idea of building summer cabins for tourists who wanted to escape from it all. On the other hand he thought perhaps they, too, should think about moving a little closer to the outside world. Sooner or later, one has to face up to the fact that he cannot be entirely independent."[2]

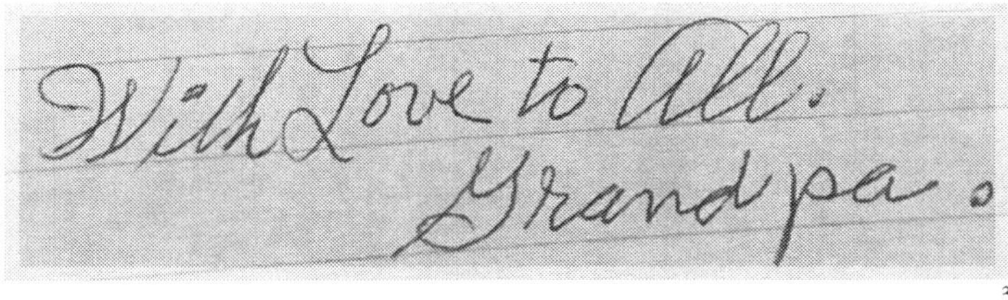

[3]

---

[1] "'I Remember Mama' Plays to Full House," *Nanaimo Daily News*, March 16, 1962

[2] R. Bruce Scott, "Independent People," *The Daily Colonist*, February 9, 1969

[3] Letter to Leona Dolling, in Nelson's last year, around 1997.

Nelson receives the line from the *Lady Rose*. Photo by Heather Arnott.

14. Tsunami

The earthquake and tsunami of the year 1700 was the last seismic "Big One" to hit Vancouver Island, and it devastated many First Nations villages along the coast. But within many people's memory was the tsunami created by the Alaskan earthquake on March 27, 1964. Amazingly, no lives were lost on the West Coast of Vancouver Island this time, but the property damage was extensive.

Nelson II was living with Earl and Louise Johnson near the Somass River in Port Alberni at the time. On his Easter break, he was visiting Copper Island with a cousin from Washington State, who drove the two of them to Bamfield where Nelson and the *Raven* picked them up. We don't know if the Dunkins had any warning by radio, but for whatever reason, Mina woke up at midnight to watch the events that would happen that night, and she stirred her husband out of bed too. Madge, Nelson II and their cousin slept through the whole thing.

The first wave hit Port Alberni shortly after midnight on March 28. It had travelled across the Pacific from the 9.2 earthquake south of Anchorage at 720 km/hour. Each wave took ten minutes to travel the 60 km from Bamfield to Port Alberni, a three-hour boat ride away. The second wave—and the largest—was one foot high at Bamfield but had grown to eight feet as it twisted up the narrow channel and arrived at Port Alberni at 1:20 AM. There were successively smaller third and fourth waves at 3:50 and 4:15 AM, plus two more later.

Fortunately, the waves came as huge tide surges and not as cresting waves. But the sudden rise of an extra eight feet of ocean during a high tide made cars float down Port Alberni roads like boats. Earl was up the coast on *Messenger III*, but his wife Louise and children were at home and escaped just in time. Their home was carried up the river and dumped on its side. With the help of friends and the grace of God, they were able to rebuild soon after—but on higher ground.

Earl Johnson remembers, "We were fast asleep in our galley bunks aboard MV *Messenger III* moored at the Esperanza airlines float. The atmosphere was quiet, the waters calm." Friends on shore alerted them to the earthquake and the waves coming their way. "In no time flat, our lines were loosed and *Messenger* moved away from the float into Hecate Channel separating Nootka Island from Vancouver Island… Floating, *Messenger* was carried east through Hecate Channel toward Tahsis Narrows as if we were on a fast-flowing river. Dead red snappers were appearing on the surface all around us. At 9:30 AM, I was finally able to locate Louise by phone. I asked, 'Should I come home immediately or finish my work in the Nootka Sound area first?' (Think what you might, my friend, your harshest judgement of me would be lenient!) Selfless, gracious Louise simply replied, 'It would be good if you could come.'"[1]

Mina, watching out the window of their new Copper Island home, saw the water quickly recede in front of their house until the boats were sitting on the ground. She could see the entire bottom of the bay. Then the tide rose rapidly, higher and higher, until the water was slopping up through the floorboards of the first storey. Nelson was afraid the flood tide might lift the house right off its pilings. But the levels quickly dropped to normal again.

This happened several times, all night long, with the levels lowering each time as the tsunami waves and the regular tide receded. There was no great damage to their boats or house, apart from a bit of mopping up to do. However, several items were washed away from the property and never recovered.

Down the inlet in Port Alberni, the situation was much more dire. Newspapers were filled with stories of destruction. Homes, businesses, boats, cars—but no lives—were lost that night. A tsunami of this magnitude was a surprise to everyone, and it took many months to recover and rebuild. Insurance companies immediately wrote this off as an "act of God," passing the buck as smoothly as Adam and Eve.

[2]

Even Camp Ross was affected: "At Bamfield, 35 miles west near the mouth of the inlet, Alfred Logan, a crew member of a lifeboat, reported 50 teenagers stranded at nearby Pachena Beach. He said the lifeboat would be sent to rescue them."[3]

---

[1] Earl Johnson, *Looking Astern*, 2018

[2] "Four Waves Hit Alberni," *The Vancouver Sun*, March 28, 1964

[3] "Boats Washed Into the Bay," *Nanaimo Daily Free Press*, March 28, 1964

View of "Temporary Cove" (as Nelson sometimes called it) from beside the old trading shed. Pencil drawing by Rick Charles, 1983. rickcharles81@gmail.com

Nelson's tool shed was probably the only building on the property when the family first moved there. It is variously remembered as one of the original trading post buildings or a supply hut used by the miners on the mountain—or both. You can see that Nelson employed his unique method of shingling on the roof.

15. Camp Ross

While I was attending Bible college at the age of 19, I heard of a camp on the west coast of Vancouver Island offering an interesting summer job. Camp Ross was a ministry of the Shantymen's Christian Association and was located at Pachena Bay, at the northern terminus of the increasingly popular West Coast Trail. The camp was looking for young adults to staff the information centre at the start of the trail in a unique partnership with Pacific Rim National Park. They saw this as another opportunity to share the love of Jesus with the 5000+ hikers and visitors who passed through each summer.

Camp Ross' hut where the author gave information to hikers of the West Coast Trail

The Shantymen first held camps at Pachena Bay in 1955, a few years before the Dunkins moved to Copper Island. The camps were initially labour-intensive. They would bring *Messenger III* into Pachena Bay and row a dozen skiff-loads of gear through the surf and onto the beach: tents, cooking gear, food, a cast-iron cook stove and other equipment. They hauled everything to a large field, where tent cabins, a kitchen and a meeting area were set up. Of course, at the end of a month of week-long camps, everything had to be hauled back to *Messenger III*, often finishing up after midnight. Finally, they would boat all night up the coast to Esperanza to do another camp in Quatsino Sound.

By the time I—and another recruit named Dave—arrived at Camp Ross in the summer of 1978, we were able to drive there via rough logging road, built in 1963. In 1966, Jim Sadler and many helpers built a spacious lodge in the field, along with cabins to replace the tents. Ron and Joan McKee arrived in 1973 to begin year-round programs at the camp, including apprenticeships for aspiring missionaries, week-long Bible training and the hosting of high school and college groups, while continuing the ministry to hikers and the summer camps. The place became known as "Pachena."

Both Nelson II and Madge attended camp there. Madge remembers going as a camper, and groups from her church also held retreats at Pachena. Nelson II told me that Earl led him to the Lord at Camp Ross when he was 10 years old, so this would have been while they still lived in Kildonan. Nelson and Mina also visited Pachena occasionally. Bill Irving remembers, "We once brought them up to Bamfield and drove them over to Camp Ross for a gathering. We were having a conference or something. And Mina was sitting in the back of the van, holding onto Nelson for dear life. She wasn't accustomed to travelling in vehicles on bumpy roads."

I remember arriving at Pachena the first night. Dave and I had already become fast friends while travelling together all day, and we looked forward to a summer of adventure. We were late for dinner, but not too late for the dishes! And there were piles of dishes, with the two of us as the dishwashers. I'm sure we washed and cleaned until midnight. Later, I learned we were being watched carefully to see how we would respond to our "initiation" into the life of a missionary. The skills and attitudes I learned at Pachena that summer have lasted a lifetime. And I have never forgotten those dishes.

A regular highlight at many of Pachena's camps and youth retreats was a trip to Copper Island to visit Mina and Nelson. Brian first met the Dunkins when he was twelve. He and a group of other youth from the camp boated over with Earl in *Messenger III*. Brian recalls, "I remember steering the boat, and I remember Earl sharing the Lord with me as we were travelling." Earl led Brian to faith in Christ that summer. "And the next," chuckled Brian.

Many remember meeting Nelson and Mina through their involvement at Camp Ross. Ben went with Jack McLeman and several counsellors and campers to Copper Island. Bill Irving remembers that Nelson II and Madge were still on Copper Island when the Shantymen began taking youth there regularly. He was a camper at first, then a counsellor, and they often travelled to the island to help with projects. Nelson and Mina's example and influence were very important to him over the years he visited.

Debbie remembers, "The Christian school from Victoria would come up to Pachena, and we would take them to Nelson's and do work projects. They would cut and stack firewood and clean the house, and we'd always have a sing-song time, and they loved it. And Mina really liked having people in the home. They let us use the kitchen and do the cooking and the baking so she didn't have to, and they could just sit and enjoy a house full of people."

Heather visited with this Christian high school when she was in grade 11. "As *Lady Rose* drew up, Nook was there to give us a big doggy-scented welcome. Nelson, quietly smiling, escorted us across the maze of logs and lumber that passed for a dock. He let us explore all through the house. The girls slept upstairs, and one night, one of the girls was sleepwalking, and we stopped her just before she headed down the steep staircase.

"When we gathered in the evenings for singing and a Bible message, I remember Nelson sitting quietly in the background—in the corner by the bedroom doorway, where Mina was resting in bed. He looked so pleased that this was happening in his house. One evening, a student went outside and told his friend he wanted to be saved; another rededicated his life to God—Copper Island was a place of decisions. Nelson and Mina's home was being used by the Lord to touch young people's lives, and I believe their hearts were happy."

Pat's first time at Copper Island was with Debbie and Roy when they took MV *Lady Rose* to attend the first "annual" Christian Artists and Musicians Retreat. She remembers Rick Charles playing the violin on the roof of the big house. Roy made some beautiful sketches. Everyone could write or draw or carve or make music as they pleased, and they all gathered together for meals. Pat and Debbie—the only practical ones among all these artists—did the cooking and stayed in the Honeymoon Cottage. Debbie recalls, "It was so much fun because you did whatever you wanted, all day. Nelson really enjoyed having the place full of people."

Other groups not related to Camp Ross came to Mina and Nelson directly. A news article mentions "25 Guides who spent seven days camping on Copper Island. The Beaufort Division camp was from July 3-10… Although a fun camp, it rained for six days and the memories of 24 pairs of jeans drying, steaming over a campfire, will linger in their memories for quite some time… The hiking trip to the old iron mine and guests Mr. and Mrs. Nelson Dunkin for a campfire ceremony were highlights of the camp.[1]

Nelson writes about Don and Patty bringing over kids who were part of their Boys' Club: "It is certainly wonderful of Don to start a work to help those wayward kids and may the Lord prosper his efforts. They even have some going to church with them. Seems they most all are half orphans. When they arrive i expect there is apt to be some thieves and vandals amongst them, and that we must allow for, seeing they have no real home life."[2]

After their visit, he added, "It is encouraging to me to see such good work being done by the Boys' Club. The kids show much better behaviour than before. Now no bickering and quarreling as aforetime. It is wonderful what Jesus will do. They have a Bible Story Book and Patty reads them a story and there are questions and they all take a keen interest in making answers. All 6 kids are sleeping in the attic of Honeymoon Cottage, Wynn down below, me and Snuggles in the Hermitage, Rick in the main room of the new house, and in the new bedroom Don & Patty. Everyone will be out tomorrow, and it is as well—before the 4-ring circus drives me crazy (that is, crazier)."[3]

Harold was a regular at Pachena and participated in several dory camps in the late 1970s. The camps started at Pachena, where campers learned to row and steer the dories—which Roy built—in the surf of the bay. When campers had some experience with the boats, the staff drove them over to Port Desire in behind Bamfield and launched the dories. They rowed the 10 km to Copper Island, stopping at a rocky islet on the way. Harold also recalls his dad Jim taking campers skiing behind his boat in Nelson and Mina's bay.

I remember one evening at Pachena when no group was on site and the Camp Ross staff rowed the dories out to the mouth of the bay. We pulled into a small cove on the north shore and hiked an overgrown trail to the soft gravel of Keeha Beach. The sun was setting over the Pacific, and we wandered to where a small river flowed across the beach. It was magical. We surfed the dories back to Pachena Beach in the twilight.

I was a counsellor when we took a group of campers over for their visit to Copper Island. Three young guys and I stayed up in the attic of the big house. Sadly, Mina had passed away the year before, so I was never able to meet her. Getting to know Nelson was a highlight of that adventure-packed summer, right up there with hiking the West Coast Trail twice and moving the pigs to a new sty. I was determined right then to return to Copper Island someday to have a cup of tea and a long chat with Nelson Dunkin.

Today also [received]: a big birthday card from Camp Ross, the kind where everyone has something nice to say. They were all kind to me—no one addressed me as, "You old motheaten antique." So makes me feel real young.[4]

---

[1] "Guides & Brownies," *Alberni Valley Times*, October 12, 1973

[2] Letter to Pat Rafuse, undated, 1982

[3] Letter to Pat Rafuse, December 12, 1982

[4] Letter to Heather Arnott, January, 1978

16. Neighbours

How do you "love your neighbour" when your nearest neighbours live kilometres away by boat? Though it may seem unlikely to people who live in urban areas, residents of rural and coastal neighbourhoods tend to look after one another very dependably. They are often one another's only help and lifeline. When my wife and I spent a summer in New Zealand, we were well out of sight of our nearest neighbours. Yet we were invited to every house and often returned home to find that someone had left us vegetables, fish, a rowboat to use or a surfboard to try.

Bill Priest was one of those "near" neighbours to Nelson and Mina, living in San Mateo Bay, about eight kilometres up the Inlet. "Most people in the Barkley Sound area didn't spend much time with Nelson, including me, mostly because we were focused on our own lives. But when we heard of something that he needed to be done, we would be over there to help him out. I would help a bit with dock repair, and Peter helped with milling lumber."

Bill remembers how they met. "I was getting our fish farm started. I poked around a bit, and I could see this grey house on the beach a long way off when I got out of San Mateo Bay, heading towards Bamfield. But I never went over there. I'm disappointed I didn't go right away so I could meet the neighbours who were as crazy as I was, living out there for no reason." The first time Bill went to Copper Island was to see some friends who were staying there while Nelson was away at

Mina's funeral. "Some time after he came back, I went and introduced myself. And by then, his house had burned down.

"When I first met him, it was in this new house that some of the locals helped put up for him. Just a simple cabin. Nelson always enjoyed whoever was able to come and visit. Didn't matter who it was, he was always welcoming and glad to have company. That's why I made an effort to get over as much as I could. I was focused on the fish farm activities, and it was a 10- to 15-minute boat ride each way, so I didn't get over there as much as maybe I should have. When I did go over, if it wasn't a social visit, there was always some chore to do."

Bill's brother Steve also loved to visit Nelson. "Nelson was a one-of-a-kind friend and Christian brother. We had numerous visits together. I played his pump organ many times, and one of his favourite hymns was 'Be Thou My Vision.' He was so appreciative of his Creator. A true pioneer in Barkley Sound. My memories of the years I spent on the coast are precious, as are the souls of those I met there."

Nelson & Nook greet Gloria & Debbie from Camp Ross, with Angie & Wilson

Nelson and Mina's closest neighbour was Angie Joe, who lived with her son Wilson and sister Eunice a short way up the Sarita River, directly across from Copper Island. Nelson wrote to Pat, "A while ago Angie and Wilson were over. Brought a pie and their mini chainsaw. We all had soup and ate the pie and Wynn fixed the crooked wire control to their carburetor." Patty remembers going over to visit Angie as well. "We helped them pick grasses so they could weave baskets. Eunice made us a beautiful little woven basket for our wedding rings. We rowed over there a few times, or we took the troller in, anchored, and rowed up the river where their house was, up by the remnant of the native village. They never had electricity, but they always baked these huge loaves of bread. Whenever we went to visit, they gave us big slices of bread and butter with jam, and the teapot was always ready."

Brian remembers that Nelson was well-connected with the people of Alberni Inlet and Barkley Sound. He sometimes took the Dunkins' laundry to town. "Of course, I'm coming in to deliver fish, so I'd stop off at Copper Island and go to town, then turn around and come back. Or if I was at Bamfield working and needed to go to Port Alberni, I'd stop by there. I might stay overnight and leave early in the morning. Sometimes, I gave Mina a ride to town. This gave her the opportunity to do things that she needed to do. Or they would give me a list, and I would get their groceries." He was also one of the several people who say they regularly gave Nelson a haircut. Bill Priest remembers, "It was interesting to see Nelson cleaned up and groomed a couple of times a year."

Nelson's most faithful neighbour, especially after Mina's passing, was Mary Scholey, a Pentecostal missionary and the postmistress in Bamfield. She deserves her own chapter, so for now I will say that rain or shine—and mostly rain—Mary faithfully brought mail, supplies and tasty snacks to several people in Barkley Sound, including Nelson Dunkin and Angie Joe. If at all possible, she did this once a week for many years, having several adventures on the way in her small outboard.

While Mina was still with him, Nelson was more motivated to leave the island and take her to town, which usually meant Bamfield, 12 km away. Nelson always spelled it "Banfield" in his letters, knowing that the town's name was misspelled "Bamfield" on the original post office's cancellation stamp. R. Bruce Scott, who lived there many years, wrote, "Twice a week, on Tuesdays and Thursdays, Nelson makes the trip to Bamfield in his commercial fishing type boat *Raven*, to collect mail and purchase provisions. Should any freight be consigned to him, the mailboat *Lady Rose* would deliver it to his own float at Clifton Point, and, should they want the *Lady Rose* to call for passengers, or any other reason, all they have to do is flag her down and she would obligingly veer from her course."

Bill Irving remembers that Nelson was dedicated to the weekly trip. "When we were down there staying with them for a few weeks at a time, there was a regular Bamfield trip once or twice a week. In the old house just off the kitchen was a huge pantry, so he really didn't need to go every week. But that was important to him."

In addition to visiting by boat, Nelson and Mina also talked with their neighbours by citizen's band radio transceiver, under the handle "Copper Island." Mina especially enjoyed these conversations with her distant friends, often chatting off and on all day. Saturday nights were a time when they could count on many neighbours being near the radio. After Mina's passing, Nelson grew quiet. He eventually gave away his electric generator, and the radio went silent.

That was the direction things took in the second half of Nelson's life on Copper Island—the trips to Bamfield ceased, he narrowed his correspondence to a few, and he more and more depended on people to come to him if they wanted to visit. But this didn't discourage them from visiting. Perhaps his isolation and reclusion were part of the draw for many visitors to Nelson's home on Copper Island.

17. Visitors: Part One

Our second visit to Nelson came as quite a shock to him. My wife Sarah and I were living in Campbell River at the time, and Don—who attended the church where I was a youth worker—was serving as a pilot for the Shantymen. One day, in the early spring of 1982, he told us that he was flying into Barkley Sound to visit an elderly fellow named Nelson Dunkin, and did we want to come? Of course we did! We jumped at the opportunity.

The three of us took off from the small airport in his amphibious float plane. We kept to the east of the mountains, flying over what looked like patches of forest between vast clearcuts. As we neared Copper Island, Don could see that the wind was averse to us landing in the bay in front of Nelson's house. We landed at Pebble Beach instead, pulled the plane up on the shore, and walked the rough trail through the woods to Nelson's house.

We knocked on the door, and when Nelson answered—well, you have never seen a more surprised look on someone's face. He hadn't seen anyone pull up at his float, and yet there we were. "How did you get here?" he exclaimed. "Are you staying for a week?" We enjoyed chatting with him over tea in the kitchen, though he was clearly disappointed that we could only stay a short time. Don was nervous about the floatplane on the beach and soon left to check on it.

Of course, Nelson probably didn't remember me among the many campers and counsellors who had visited him from Camp Ross several years before. But he was very pleased to see me and to meet Sarah. "You must come back soon and stay for a week," he insisted. We promised we would. By then, Don had taxied the plane around Clifton Point to the float. Nelson came down to see us off. On the flight home, as we buzzed over Mt. Washington to watch the skiers, we hit an air pocket and dropped 100 feet. I said to Don, "I'm sure that didn't even faze you." He replied, "Actually, that scared me too!"

Winters must have been lonely for Nelson after Mina was gone. But as soon as the weather warmed, he could count on more visitors arriving. Despite his policy that guests didn't need to make a reservation and should just show up, his calendar was often full of names of expected guests. Ben says that Nelson was irked by his son-in-law Andy's frequent suggestion that he move into town so he could receive more visitors. "One summer day, when I was visiting Nelson, he asked me, 'So, Ben, how many visitors do you think Andy had this month?' I replied, 'Probably not many.' With a twinkle in his eye, Nelson said, 'I've had 42 visitors this month.' It shows how beloved Nelson was to many people."

1

Nelson and Snuggles looking to see who has come to visit.

Inger remembers that Nelson loved it when sailboats or kayakers would stop in for a visit. An article on him reads, "He is perhaps best remembered for the hospitality of his unique wilderness home on Copper Island in Barkley Sound. With his wife Mina, Dunkin welcomed fishermen, hippies, youth groups, or anyone else who stopped by. 'Always their place was open and welcoming,' says Brian Burkholder, part of the Coastal Missions team. 'They were happy to see a group of twenty or two. And they treated you like you were the most important person in the world.'"[2]

Roy told me, "I always remember Nelson coming down on the floats—and being a bit shaky—but very welcoming when people arrived there. Nelson never wanted you to leave; it was always a struggle when you needed to go, to get yourself back to Bamfield with enough time. So on the one hand, he reclused himself from people, but on the other side of the coin, he never wanted people to leave."

Well, almost never. Ben mentioned that Nelson had several signs with slogans on them around the place. One read, "All of my visitors make me happy: some when they come, others when they go." Indeed, some visitors were not easy on Nelson, no matter their good intentions. Heather remembers helping a youth group that set out to burn up some old mattresses. However, the fire went underground and wouldn't go out. "We took turns all night watching it. One lady began to swing a lantern, seeking to alert a ship, and Nelson stopped her, not ready to attract that kind of attention. The fire did eventually burn itself out. Nelson later recommended me for 'missionary training credits' for 'combatting a fire which but for the combined efforts of all involved and the mercy of God would have developed into a forest fire of considerable magnitude.'"[3]

Several people recounted times when Nelson would quietly slip away from his guests. Patty told me that when Nelson was living in the Honeymoon Cottage while his new house was being built, he moved out and stayed in the tiny "Ye Hermitage" because the fellow staying with him wouldn't stop talking. Bill Priest thought that Nelson "needed to escape to a place where he could have his own space and not be with the guests who were there at the time. But he was very thoughtful and always put other people ahead of himself."

"Ye Hermitage" was along the west bank of Nelson's bay and was once an outhouse.

Nelson wrote, "What does irk me are the Christians supposedly coming from good homes who never put a thing back where it belongs, never return what they borrow, have no conception what-so-ever what a broom is made for, and if Bibles or hymn books fall on the floor they just walk over and kick them about. i say these things in pain of being branded intolerant. i like a homey home which one can live in—not something like the inside of a cold storage plant. Still, i like to have a place for everything and everything in its place."[4]

I'm afraid I have to put myself in the category of not always helpful. I had hiked up the mountain to the mine shaft and, poking around, saw what I thought was some interesting mining equipment. When I told Nelson about what I had found, he wanted to see for himself. I protested the idea of him making the trek up the mountain with his health at the age of 75, but he insisted. I wish now I hadn't let him, especially when what I had discovered turned out to be of no use to him. Though he was exhausted by the time we returned, he never said a word of complaint and only talked about how long it had been since he last saw the old mine.

A few notes from Nelson's letters about visitors:

Thursday now and Sunshine: [Ben] Potter & Co. out fishing and will be going back to town when they come in so will take this letter to post. They invite me to eat with them which i do not refuse. Hope to get more done on the canoe. Still looking for my son and family.[5]

Brian, Ann, Tom & Debbie were here and stayed overnight, next day going on to Kildonan. They had me eating with them and Debbie baked a chocolate cake for me. They told me all about the girl in a wheelchair who has come to help them in the office work. She is quite a wonderful person. Wouldn't it be nice if i could do something for the likes of her?[6]

Labour Day keeps me looking for company: Expecting two families from Seattle and Olympia. i sure need a receptionist here. My niece was here from Centralia—four of them and one baby. She gave me a camera so i am back in the photo business again—a very expensive hobby. David & Arlene of Kildonan were visiting. Well anyway some of my many guests should be arriving tomorrow. Lady Rose heading this way.[7]

Be sure next time you come to bring your fiddle along as i have over 1000 jigs, reels, hornpipes, long dances and marches for fiddle in O'Neill's Music of Ireland which Ron picked up on the street in Victoria. Bill Priest was in to visit and left a bag of cooked salmon. Now it remains to wait and see if Mary will come tomorrow. PS – My hair was so long that it would tie a pony-tail, and Patricia cut it. Jim & Dodie were here last Friday and brought cookies, nice big Yum Yum cookies which Dodie had made.[8]

Pastel by local author and artist, Lynn Starter

---

[1] Bamfield Community Museum and Archives

[2] Adele Wickett, "Remembering Nelson Dunkin," *Island Christian Info*, June 1998

[3] Sent to Heather Arnott, August 29, 1979

[4] Letter to Pat Rafuse, undated

[5] Letter to Jim and Sarah Badke, August 16, 1987

[6] Letter to Jim and Sarah Badke, May 2, 1988

[7] Letter to Ron Pollock, August 23, 1986

[8] Letter to Don and Patty Cameron, January 3, possibly 1981

18. Visitors: Part Two

Most of Nelson's visitors made a point of being helpful and generous. In a letter to us, he says, "i was blessed to have Don & Pattie and their 3 darlings come to visit me. And here comes Ron on the *Lady Rose* with a whole tray of freight, including a bathtub. So if God be pleased to heal me there is much for me to do. The Toy Factory, the windmill-boat and some exclusive rustic cabins over and beyond the waterfall… Well, my friends have just left for Port Alberni as the weather is so cold and wet, there is no pleasure in staying. They go back empty and leave me with eggs, milk, sausages and chicken.

"Yesterday a man was in to see me, a Christian logger, and he has a restaurant in Tofino. He knows just about everyone whom i know. So he was a great encouragement to me. Being so interested in this place, he would like to come here to work at such time as the Lord will straighten his business out. Went with him over to Pebble Beach which he thought was ultimate. He also thought the water-power idea excellent. And the Toy Factory. i felt just fine but perhaps i shouldn't have made the trip as last night i seem to have relapsed back into some of my old pains."[1]

Debbie recalls, "One year we went and stayed there and cleaned his place as best we could in the four or five hours that Brian took Nelson to Kildonan. Another time, Gloria and I went for three or four days to be with him and do spring cleaning, and to cook and bake. And I remember we washed the windows and that kind of stuff, just so he had company.

The priority was spending time with him and then doing the work projects. He was really happy we would do that."

Several people made a point of checking in on Nelson regularly. Brian remembers that the workers at Camp Ross made sure people visited Copper Island often. Heather said, "The wood pile was ever-present, waiting to teach us how to chop and split." When Bill Irving moved to Ucluelet, he and his family and friends and his house church would travel with his boat back and forth to Copper Island pretty regularly. "The last probably ten years of visiting, I was just there to find ways to make him comfortable, like cutting wood or whatever we could to help him out."

Ben remembers, "Nelson loved to join us in the evening for fellowship, drink hot chocolate, and enjoy a generous amount of my wife's cookies. He loved those. These were a luxury for Nelson. My wife and I referred to him as the "Cookie Monster of Copper Island." Irene added, "I gave him a fairly rough time, but it was good for him."

Copper Island
Sunday December 9

Dear Ben, Irene & Little Potter :

Thanks so much for those delicious cookies i enjoyed eating what i had — the mice must have found their way to the cookie jar as they seemed to go too fast.

With Christian love
Nelson
The Copper Island Cookie Ogre.

2

Nelson's family and relatives also came to visit. As a young man, Brian regularly took his boat from Bamfield to Copper Island with Ardis Longabaugh, who was Nelson's cousin, and her husband and children. "They had a pretty girl, so I had to hang around," he laughed. The girl, Sherry, made a point of travelling to Nelson's funeral.

The children of both Madge and Nelson II enjoyed visiting Copper Island to see their grandparents. Nelson II recalls, "My father absolutely loved my children and step-children, and they loved him. He took them as his very own. He absolutely doted on the three girls, especially Laura, Heather and Destiny. They used to go over, and he carved with them and drew with them, even when they were tiny little girls. They spent many hours with him when he lived in the house. He absolutely adored them, and they adored him."

But there were long periods when Nelson had no visitors, especially in the cold and wet months. He wrote to Ron, "I keep busy but seem to not accomplish very much. And with the weather what it is, Bill don't get over like in summer, neither can Mary come every week as is her custom, so I just take my company as they manage to get here and thank the Lord for them. Jim has not been ever since they got married—of course he is out of the hermit class now.[3]

Havn't seen Wynn since the 5th this month.
" " Bill Priest " " 20th last "
" " Mary " " 15th " " .
" " Jim " " Bible Study.[4]

This last month, i had 99 people coming and going and some overnighters. So that was most encouraging to me.[5]

---

[1] Letter to Jim and Sarah Badke, May 1988.

[2] Card sent to Ben and Irene Potter, December 9, 1983

[3] Letter to Ron Pollock, November 20, 1984

[4] Letter to Ron Pollock, July 22, 1984

[5] Letter to Ron Pollock, September 1, 1985

19. Visitors: Part Three

My wife Sarah and I kept our promise to come back and visit Nelson for a week—not once, but four times. Our first visit was soon after the original house burned down and before the new one was built. I remember making the precarious leap from the hull door of the *Lady Rose* to his float as it surged up and down. Nelson and Nook were there to greet us. He was living in the Honeymoon Cottage at the time.

One morning, a boat pulled up and two young men asked if they could moor there for the night, as they planned to do some scuba diving in the area. Nelson, of course, gave them his blessing, and they in turn offered to provide dinner for us all. They came back hours later with a large bucket of abalone (which was still fair game until all northern abalone fisheries were closed in BC in 1990 due to conservation concerns). Our new friends showed us how to tenderize the strange-looking things with one of Nelson's mallets. They fried them up in bacon fat, along with potatoes—and I seem to remember corn on the cob. To this day, I have never tasted anything better than that abalone. Like a giant scallop, it has incredible flavour and just melts in your mouth. I can say our best meal ever was on Copper Island.

We had a busy schedule while I attended seminary, so we didn't return to Copper Island for three years. By the spring of 1985, Nelson's new house was complete and already looking very lived-in. I remember we took every opportunity—while Nelson was doing something in his shop below—to clean, clean, clean. I felt torn between wanting to help him in any way I could and Nelson wanting me to work on my personal projects instead. I would have been fine with gathering firewood or repairing his float. He wanted me to try his wood lathe or scavenge wood on the beach for my carvings.

We didn't know until later that Sarah was expecting our first child on this trip. By the time we returned next summer, I had graduated from seminary, and our son was already eight months old. Nelson wrote in a letter to Ron, "i manage to keep busy every day. Jim & Sarah and baby Benjamin Badke were here for a few days. Jim laid out a new road to Pebble Beach and did considerable work on it." I had been concerned about Nelson navigating the old trail to the beach, which was steep and very overgrown in places, and it passed through a marshy area. I blazed a new trail with a gentler, drier slope, up and over the hill to the beach. That path, much improved, is the one still used by the camp today.

Over the winter, Nelson had been working on a new project. "So as i said i am quite busy these days," he wrote to Ron the previous fall. "i am building the 'Empress Room' on the house so when you and Pat come again to visit you can have a comfortable room. Mary has ordered from Sears the best queen-size mattress so you can sleep feet inside the bed.

The sawyers haven't been able to come from Kildonan yet, so much of my lumber i have to split from cedar." Later he wrote, "The Empress Room is ready for guests. And Believe It or Not, so far i have kept myself from loading it with junk—even Snuggles and Hawkshaw are not allowed in."[1]

The author's wife Sarah and 8-month-old son Ben in the new Empress Room, 1986

The room was built out from the east wall of the second story like a large bow window. Cedar-lined, with carvings, of course, and just large enough for the queen-size bed, which had a couple of steps up to it as it was quite high. The space was perfect for us and our baby son. We felt like we were staying in a fancy hotel—at least in comparison to many of the accommodations on Copper Island! On that trip, we met the famous Mary as she paid her weekly visit to bring mail, ice cream, friendship and conversation. Nelson obviously enjoyed Mary's company—and the ice cream—very much.

We returned the next spring for what we didn't know would be our last visit to Nelson on Copper Island. Our son Ben was now a toddler and absolutely loved Pebble Beach, small crabs being one of his favourite things on Earth at the time. Nelson kept me busy in the shop, making a candlestick on the lathe, while Sarah did more of the cleaning as Ben napped. I made my infamous trek up to the mine with Nelson. We noted that he did not seem as well as the summer before.

Much of Nelson's conversation with us that year was about his dream for "God's Property," of which he considered himself the steward. It was obvious he wanted us to come and live there with him. And others: "Am trying to interest Mary in the Copper Island Enterprises," he wrote us, "as she is a good worker and would be an asset to the Company."[2] We felt torn. On the one hand, we were free—I had been searching all year for a place of ministry after seminary. On the other hand, we had recently agreed to spend the summer at Camp Qwanoes to lead their Counsellor-In-Training program. We also hoped to be called to a church in the fall where I could serve as a youth worker. We had a toddler now and much responsibility, and our resources were depleted after schooling. We didn't feel up to the expectations Nelson clearly had of us. And most of all, God was telling our hearts that this opportunity—though tempting—was not for us. We didn't tell Nelson at the time, but we knew we would never live with him on Copper Island.

On our last morning, we went to Pebble Beach to let Ben explore, knowing that in the early afternoon, *Lady Rose* would pick us up on her way back from Bamfield. It was a beautiful, sunny day, as the island often receives in the spring, but our hearts were heavy at the thought of leaving. Suddenly, we heard the loud whistle of the *Lady Rose*! We jumped to our feet, scooped up Ben and ran toward Nelson's float. It turned out that the boat would be passing by the far side of Copper Island on its trip back that day, after dropping off kayakers in the Broken Group Islands, so they needed to pick us up on the way to Bamfield instead of on the way back.

We quickly tossed our bags on board, gave a far-too-quick hug and farewell to Nelson, and steamed out of the bay and out of sight. We would never see him on the island again, and we would not return there ourselves for 29 years.

The author turns a wooden candlestick on Nelson's foot-powered lathe.

---

[1] Letter to Ron Pollock, October 6, 1985

[2] Letter to Jim and Sarah Badke, October 1987

Gifted in the art of listening with genuine interest and understanding, Nelson was quick to set down his tools and put the kettle on for tea.
Photo: Bamfield Community Museum and Archives

20. A Gaelic Visitor

by Margaret Stewart

During the summer of 1975, I came from the Outer Hebrides of Scotland for an extended holiday on the West Coast of Canada, staying with various relatives and friends. I had just left high school and had rarely left my protective island environment, so it was a huge adventure for me. My most unforgettable experience was when my MacIver relatives took me to visit their cousin Mina, who lived on a small island with her husband, Nelson Dunkin. Mina had met Nelson when he stayed at the house of a near neighbour, Mrs. Isabella Paul, while on leave during the Second World War.

Having no idea what was ahead of me, we travelled by ferry from Horseshoe Bay and then drove to Port Alberni. We may have stayed with Mina and Nelson's son or daughter in Port Alberni overnight before continuing our journey the following day along the length of the Alberni Inlet on the *MV Lady Rose*. It was a beautiful sunny day, and proved to be a spectacular and awe-inspiring journey. I still get a tingle in my spine when I think of it. I was particularly delighted to find out recently that the *Lady Rose* was a Scottish, Clyde-built vessel, the first diesel-powered ship to cross the Atlantic by a single propeller.

We stopped at various points along the inlet to drop off and take on passengers and freight. It was a longer journey than I expected, but I was in my element as I loved boats and being on the water. There was a film crew on board too, although I don't know for whom they were filming. It was an interesting coincidence that, a few days later, my cousin

Wendy in Vancouver happened to be watching TV and saw me on the *Lady Rose*.

As we approached Copper Island, I can clearly remember seeing this huge cabin, built on wooden piles on the shoreline. I don't know what I was expecting, but it certainly wasn't an enormous house. For one who had been brought up on a treeless island, it all seemed like a perfectly-staged movie scene. I alighted precariously from the shell-door of the iconic *MV Lady Rose* onto the float. Mina and Nelson were expecting my cousins, but I was a complete surprise (possibly in more ways than one, being dressed, *à la mode*, in tartan-trimmed Bay City Roller style denims). During her life on the island, Mina had only seen two or three of her own people, so when I greeted her in our native Gaelic tongue, she was quite overcome with emotion.

The cabin was a joy to behold. It had three floors, and when you stepped inside you were enveloped in the beautiful scent of cedarwood. The place was full of Nelson's carvings, mostly of biblical scenes, and there was a fabulously carved cradle. When the tide came in, I could hear it lapping against the cabin infrastructure. It was like nothing I had experienced before. I explored the island as much as the dense forest and shoreline would let me. I chatted with Mina in our native Gaelic tongue and tried to answer her many enquiries about people back home. My memory of Nelson is of a quiet and gentle presence with a lovely twinkle in his smiling eyes.

Life on Copper Island seemed to me an enviable existence. But to one such as Mina, who had been accustomed to a large and close-knit community in her native Coll, I'm sure it felt lonely at times, especially during the winter months. Nelson and Mina's house was obviously built to accommodate more than just the two of them. It became apparent that they were used to welcoming many visitors, particularly during the summer months.

20. A Gaelic Visitor

I was sitting out on the float one day, dangling my feet in the water in the manner of Huckleberry Finn, rather wishing I had younger cousins or friends with me to share this experience. Around the headland came a dory, crewed by a group of young people as if sent to grant my wish. While I enjoyed being with my relatives, who were around my parent's age, and loved chatting to Mina about Lewis, this group from Camp Ross at Pachena Bay brought an extra dimension to my visit. They stayed with us for a couple of days, helping with chores around the cabin site and taking part in prayer sessions. Having been brought up in a very religious community, this was not new to me, but I did find it unusual that such young people were so steeped in the doctrine.

I don't know who suggested it or why—perhaps they sensed that I needed the company of people my own age—but our visitors invited me to return with them to their camp for a few days. When I tell folk about it now, they often think it was a crazy thing to do. How did I know that I wasn't being kidnapped to join a cult? But I am sure the adult cousins wouldn't have permitted me to go if they didn't think it was safe. Off I went anyway. We rowed for quite some time until I began to get blisters on my hands, having not been accustomed to rowing for any great distance. We stopped off at a beach on a very small but lovely sandy island to have lunch and a short break. As we continued on to Bamfield, they let me take a rest from rowing. I eventually fell asleep in the bow—assisted, no doubt, by the gentle swell. They were singing, and at one point they sang "Speed Bonny Boat," the Skye boat song from Scotland.

We were then driven to beautiful Pachena Bay, where I met the rest of the people at the Bible camp. I was treated with great kindness. I remember feeling at home when down at the beach, as it reminded me a bit of Coll Sands on the Isle of Lewis. I enjoyed visiting Anacla, the newly re-established

Huu-ay-aht village next door, and seeing their ceremonial canoe. The whole experience at Pachena Bay was a pleasure. I corresponded with someone from the camp for a while, though we eventually lost touch.

Campers in a Pachena dory, built by Roy Getman.

At the end of my visit, I was taken back to Bamfield to catch the *Lady Rose* back to Copper Island. It was a great privilege to spend time with Mina and Nelson in that very special place, especially since Mina sadly passed away only two years after I visited them.

The director of the current Copper Island Camp told me that everyone who visits is given information about Nelson and Mina. "We have a plaque, pictures, a portrait and some of his special carvings on display at the camp. We honour their desire to have a Bible camp for the First Nations youth of the area. Anyone who visits is informed about their life, their love of God and their love for people."

It is reassuring to know that Nelson and Mina's legacy continues to live on in that beautiful place. I often think about Copper Island and occasionally visit it on Google Earth. I like to think that I'll manage a return trip before age gets the better of me.

21. MV *Messenger III*

"The waves were 20 feet high, and the *Messenger III* rolled and bucked as it tried to sneak sideways in between two jagged reefs off Vancouver Island in treacherous waters called the 'Graveyard of the Pacific.' Aboard were three missionaries from the Shantymen's Christian Association, a non-denominational organization in Canada dedicated to bringing the gospel to people who live—mostly in shanties or rude houses—in lonely and isolated places."[1] This description was written by a *Life Magazine* journalist who came along for the ride in 1954.

He caught this shot of *Messenger III* in heavy seas, Christmas tree tied to the mast.[2]

The journey of a Shantymen missionary on the West Coast was one of persistence and endurance. "If you have the zeal of a Billy Graham, the toughness of a Green Beret, the desire to lead an outdoor life in some of the most rugged terrain in Canada and, in the time of inflation, can live on about $35 a week and expenses, then consider joining the Shantymen."[3] Sarah and I had the privilege of talking with one of the legendary Shantymen who travelled the Coast on *Messenger III*, Earl Johnson.

Earl came to our table at White Spot in Campbell River while he chatted up the server in a very friendly manner. He was glad to meet us and was more interested hearing about the two of us than talking about himself. "Well, I'm here to learn about Nelson from you!" he said. We were amazed at Earl's energy and drive. "At 94, I'm still 16 days of the month away from Campbell River, caring for the hurting. But the Lord is good." He told us that last Easter, he had decided to visit friends in Alberta. He jumped in his car after a family dinner and drove all night and the next day, 36 hours, to Three Hills.

Earl and his fellow Shantymen spent time with the Dunkin family in Kildonan and on Copper Island. "We visited Nelson with *Messenger III* for 20 years, and I was there for 16 of those years," Earl remembered. "Nelson was meek, wonderful, gracious. Very knowledgeable in life. And he was a quiet soul, as you know. I don't know if I'd call him a recluse. I visited the recluses on the coast. Mina was outspoken, straight from the shoulder. She didn't mince words, nor waste words. She was a kindly Scottish soul. Mina was strict, but she was a fun lady, too."

I asked Earl what a visit with Nelson and Mina was like. Though he couldn't remember the specifics, he told us, "Well, Shantymen approach was always to come alongside and enjoy what they're doing, be a part of what they're doing. If they're splitting wood, that's what you do. You don't say, put the axe down because I've come here to talk with you. No, it's, 'Where's that other axe you've got?' And you lighten their load. You come in to see if you can leave something of the care of God, the assurances from God. So you walk in and lift loads.

"I would ask about how things were with them, as time went on. They'd have a meal or something, and we would talk about life, whatever it was, the children or the sickness, the heartache. We enjoyed the Word and prayer with Nelson and Mina. They loved to pray. Very reflective, thoughtful, deep, deep, though feisty. They were very godly, very committed.

And they loved the prayer time. They loved the word. Our visits with them were usually brief. We travelled a long coastline, back and forth, the West Coast. So maybe two, three hours would probably be a maximum, three or four, sometimes."

Nelson II and Madge remember those visits. "Oh, yeah. They would always come, all the time," Madge told me. "It was a highlight. It was good because we had people, and they were good people. Occasionally, they would take us to town." In his memoir, *Looking Astern*,[4] Earl mentions taking the family to Bamfield. Madge also remembers another mission boat that visited, the United Church's MV *Melvin Swartout*, named after a Presbyterian missionary who served on the West Coast.

Nelson II recalls his father helping the Shantymen with *Messenger III*. "We were very involved with them. When I went to the Kildonan School, Earl Johnson came by. He was a brand-new missionary. I was in grade two. I was sitting there having lunch. He walked in and introduced himself, and we became friends. At camp, when I was ten years old, he was my counsellor, and he's the one who led me to a profession of salvation."

The *Life Magazine* article highlights a visit made by Shantymen Harold Peters, Percy Wills and Earl Johnson to the Dunkin family in Kildonan at Christmas. "On this voyage, as on all their voyages, the Shantymen missionaries came not only to tell the story of Christ but to demonstrate the spirit of Christianity. Where it was needed they gave away bread and fish and clothing, asking nothing in return. At Dunkin Bay, one of the wettest spots in the world with an expected 400 inches of rainfall this year, they anchored on a rainswept night to bring a tree to the children of Fisheries Guardian Nelson Dunkin, who live on a tiny 'floathouse.'"

"First Christmas tree ever had by the Dunkin children is decorated on *Messenger III* with missionary's help. Their own houseboat is too small for tree."[5]

In his book, *Looking Astern*, Earl tells of another Christmas adventure, this time to Copper Island. He retold the story to us: "It was Christmas time, and we were going to have a teen retreat at Camp Ross. But with two feet of snow, the team gathered in Port Alberni, ten of us, and we cancelled the retreat. So I took the leaders for a four-day outing in the Barkley Sound area. We were on our way down the canal, and I heard Estevan Point radio talking to the RCMP Ganges. A boat had gone down with five people aboard in the Alberni Canal area. They couldn't hear each other on the radio, and I guess I was in between. So I called Estevan Radio, and I said, 'I hear your attempts. Could I relay a message?' I entered into the search and agreed to follow the starboard shore down as far as Cape Beale among the islands out there. As we returned, we stopped in at Bamfield to visit with one of the homes, and the MV *Duncan Scott*, the Indian Affairs boat, was there.

"We came back up the channel to spend the night at Copper Island. The girls were going to go ashore, and the men would stay on the boat. So we're all gabbing in the wheelhouse, having fun. And coming up to Copper Island, I stepped out to the door of the *Messenger* and looked out. I saw the light of the house, and I reckoned I was past the point. It was midnight, pitch black. So I swung in to go into the bay, and I was talking too much. I should have seen the light the whole time, but you know, some of us aren't as smart as we should be. So all of a sudden, one of the gals in the wheelhouse says, 'What's that? What's that?'

"And it all happened in split seconds. I thought, well, I could put the spotlight on and let her know we're not close to the float yet. Or I could wait and she would soon see. I flicked the light on, and we were headed straight toward the rocks at the point. I throttled back and threw it hard to port—that's where the ship turned fastest, to port—and waited to see if our stern was going to hit the beach. And we cleared the beach. And I said, 'Thank you, God.'

"And I said to the little lady, 'What did you see?' She said, 'I saw the anchor in the mast light.' She hadn't noticed the anchor before and wondered what it was. We would have hit head-on. Dr. McLean always said, 'The Lord always looks after the drunk and the missionary,' so how good the Lord has been to us. So every once in a while, Jim, my heart just silences. And I shake my head, and I say, 'Oh God, how good you have been.'

"But that's crazy. A skipper, a skipper! I had my master's papers, and I'm doing a fool thing like that. But you get caught up with the fun and enjoyment of six or eight people in the wheelhouse, all talking about what we've been doing or what's happening. So the Lord says, 'The captain isn't on the job. So I'll prompt a little lady.'"

Bill Irving recounts another close call with *Messenger III*: "Earl would go down to Copper Island quite a bit in the *Messenger*. And before the Shantymen sold the *Messenger*, I was the deckhand with him on the boat for a period of time. We were down at Copper Island and leaving with a bunch of kids. Nelson was down on the boomsticks saying goodbye, Earl was up on the top of the boat and Mina was up on the dock. I thought I was putting it in reverse to back out. And I was actually pushing the forward button. So the boat kept getting closer and closer to Nelson, who was standing on the logs.

"Earl and Nelson both saw what was happening. Earl rushed into the cabin, 'Bill, what are you doing, what are you doing?' And I looked up, and Nelson, he's smiling. He's just standing there. If he fell in the water, whatever. It's only water. But there were the two different personalities: the supercharged Earl and the laid-back Nelson. Very interesting to see the Christian dynamic. It's good for young people to see that there's all sorts of people in God's family. We never knocked him off the logs. He couldn't swim, though he had been in the water a few times. I remember navigating the boomsticks. It was often the fun part—him wandering on the boomsticks with his boots on and us racing along with our running shoes. It was always a challenge to keep up with him."

Ben said, "Nelson never did learn to swim. He fell in one day off the dock. So he dog-paddles around and manages to keep his head above the water by thrashing. And he pulls himself up against the dock or at the shore, one of the two. But he did make it out. Nobody else around; could have drowned. But again, it didn't stop him or make him say, I'm going to have to leave this doggone place. I'm too old for it. No, he philosophically carries on. He wasn't a quitter."

Brian relates a story that demonstrates the unique humour of the coastal people. "Percy Wills and Harold Peters had stopped by Copper Island to have a visit, and they were invited for dinner. And there's *Messenger III* tied up at the float, and they say, Mina, we'll look after the dishes. So they got busy in the kitchen—but they put all the dirty dishes in the oven of the wood stove and left. They were well remembered for that! Who knows what the joke was all about? But that was what they did."

Shipmate Wilbur with Earl Johnson on Messenger III

R. Bruce Scott summed up the work of the Shantymen in 1974: "Barkley Sound still contains a few rugged characters who preserve their individuality by being as independent as possible of society. Forty years ago, there were a number of these "shantymen" or hermits who had renounced civilization for one reason or another—possibly to escape the law, or a wife, or they were just plain fed up with society. They lived off the land, hunting and fishing and spending most of their time with the fundamental aspects of life.

"To serve these men was the original aim of the Christian Shantymen's Association, whose vessel the *Messenger*, manned by the Reverend Percy Wills of Victoria, Harold Peters and Earl Johnson, plied the waters of the west coast calling on all known shantymen in the bays, inlets and islands of the coast. They gave them magazines and books and, if so desired, held a service on the boat.

"Nowadays there are few, if any, of these individuals, possibly because one can insulate and isolate himself as effectively in the modern city—and live on welfare at the same time. Perhaps this was the reason the *Messenger III*, so well-known on the west coast, was recently sold."[6]

[1] "The Mission of the Shantymen," *Life Magazine*, January 11, 1954. Available at archive.org by searching "Life Vol. 36 No. 2"

[2] "The Mission of the Shantymen," *Life Magazine*, January 11, 1954.

[3] *Toronto Star*, 1970's, scainternational.org/history

[4] Earl Johnson, *Looking Astern*, 2018

[5] "The Mission of the Shantymen," *Life Magazine*, January 11, 1954

[6] R. Bruce Scott, "Independent People," *The Daily Colonist*, February 9, 1989

22. Grace and Graciousness

If you have read this far, you have seen that Nelson had a difficult start in life, and later, he at times made life difficult for his family. Yet when people describe Nelson to me, they do so in fond and glowing terms: a gentle man who could do no wrong. When I observe the incongruity in this, it speaks to me of grace. Mercy is not getting the consequence we deserve, such as a punishment; grace is getting the benefit we do not deserve, like a gift.

When my dad passed away, I saw in his funeral plan a favourite Bible verse: "And God is able to make all grace abound to you, so that having all sufficiency in all things at all times, you may abound in every good work."[1] What God pours into us—deserved or not—we are able to pour out in good deeds toward the people around us. This describes much of my dad's life and my own (believe me, what God has given me is not deserved!), and it also well describes the life of the Dunkins. This is a chapter about the graciousness of Nelson and Mina toward (nearly) all who came to their door—graciousness that overflowed from the grace given to them.

I earlier told of Brian's visit to the Dunkins at the age of 12. As Brian grew into his teens, he wandered from his childhood faith but not from his friendship with Nelson and Mina. "The Dunkins played a special part in Brian's life. 'I knew that how I was living my life was wrong,' Brian recalls. 'I had accepted Christ as a child at Camp Ross but had gotten away from him. As a fisherman and logger in the area, I would often stop at the Dunkins' and I would bring friends to visit them.'"[2]

Brian told me, "I had my own power saw, and I cut firewood for him. And of course, when you're young, it's like, Here's a job to do; let's get it done. And so you're working as hard as you can, and Nelson would come, and he'd sit down and want to talk. He had genuine, good questions. He wanted to talk about important things. Well, I didn't have time to sit down and talk. But you had to; that was more important to him. And I had to understand—yeah, I'm not here just to do firewood; I'm here to be a friend."

"When Brian began to realize that he really needed the Lord, he cried out to him for salvation. 'Later I went to see Nelson and Mina. I thanked them for loving me through the problems and always being there.'"[3] Brian became involved at Pachena and eventually sold his fishboat. The Jim and Ruth Sadler family were also his heroes and were a great example to him.

Brian continues, "I once asked Mina, when I came back to the Lord, why they didn't tell me that I was living wrong. She told me, 'We prayed for you, and we cried for you before the Lord. But he wouldn't let us say anything about how you were living.' I had the attitude that if they wouldn't accept the way I was and what I was doing, why would I come back again? But they just loved me and my friends. They welcomed us and made it so we wanted to come back again."

Bill Irving's experience was similar. "I can't recall anybody ever having an argument with Nelson. If you weren't doing a job the way he thought it should be done, he would tend to just watch. Once you were finished, he would come along and try to help, but he probably would say, 'Hmmm, I've never done it that way before.' We often used his come-along to pull boats up or pull logs off the beach, and he would ensure that young people who didn't know a powersaw from one end to the other got instruction on how to use it, just to build up their confidence. Firewood was a by-product of having a good relationship."

This reminded Bill's wife Laurel of a story: "We went to visit him after we were married, and we'd been having a bit of a 'discussion.' I wanted to get a new dress. I remember it being a beautiful blue Georgette dress. It was $60, which was quite a lot at the time. So I said to Nelson, 'What do you think, Nelson, buy the new dress or the tools we need?' And Nelson said, 'Oh, the new dress, of course.' He had that very gentle humour. It was lovely. I always remembered that, and I think I did get the dress, on his good advice."

Donna has a childhood memory of the Dunkins: "When I was very young, my dad (Chuck Green) and his brother Doug operated Marble Cove Logging on Copper Island. One winter's stormy night—not sure now if we were coming or going to the logging floathouse at Marble Cove—anyway, Dad thought it best to hunker down at Nelson and Mina's for the night. They were very hospitable and took us in, as it was so very stormy. Mina had a meal for us, and I'll always remember the yummy homemade chocolate cake. Gosh, they were such wonderful people. I'm in my mid-sixties now, so that was around 60 or so years back."

Dan, a friend of Nelson II, was a hippie, on drugs and trying to live off the land in Barkley Sound. "Eventually, he lost his boat—a rowboat, I think it was—and it drifted in to Copper Island," Nelson II recalled. "My parents secured it and somehow found out who it belonged to, and Dan ended up staying on the island with them. My mother led him to the Lord, and he had a complete change around."

Pat said Nelson and Mina prayed long-term for everyone who came to Copper Island. "You knew Nelson was praying for whatever was God's will for your life." She also noticed that, when kids came to the island, Nelson didn't talk over them like many adults do. He loved kids and talked with them directly. Patty remembers Nelson's sense of humour while she and her husband lived with him. "Nelson had us laughing many times."

Some people tested his sense of humour to the limit. Peter remembers conversations on the radio, especially the times Nelson would talk with Clive, who lived at the Kildonan post office. "Because back in those days, nobody was around. Logging didn't start until 1971, so before that, they'd have a nice cordial chat and check in with each other. And this Jack Hobbs guy would get on the radio and go, 'The devil is alive and well in Kildonan!' Just trying to get Nelson going. And Nelson, of course, never took the bait."

At the beginning of this chapter, I spoke of Nelson and Mina's graciousness toward (nearly) all who came to their door. There were indeed some exceptions. People of "an executive frame of mind"—which to Nelson meant someone unfairly pushing their own business agenda—were not welcome. And Greg Matthews' mother told Peter of a time when someone asked Nelson how he would respond if the CBC wanted to do a story on him. His reply? "I would fetch my rifle and kindly ask them to leave."

Boat now leaving so remember i am praying for you 3. With Love Nelson[4]

---

[1] 2 Corinthians 9:8, ESV

[2] Adele Wickett, "Remembering Nelson Dunkin," *Island Christian Info*, June 1998

[3] Ibid.

[4] Letter to Jim and Sarah Badke, August 16, 1987

23. Mina

This book is about Nelson Dunkin, but I can't write about Nelson without writing about Mina as well. Though I never met Mina in person, I believe I saw her stamp on the life and character of Nelson, the man she married as a war bride in a stone hut on another island, far away from the West Coast of Vancouver Island.

"Mina, you know, such a sweetheart," Brian remembers. "How do you separate Nelson and Mina? If it wasn't for Mina, I think Nelson would have been a hermit. I mean, that was his nature. But Mina needed people, and I would say she attracted people. People came to Copper Island because of Mina. Not saying they didn't come because of Nelson. He was very mellow and very kind, and people crowded around him, but what a sweetheart Mina was. Nelson was always appreciative of people coming, but Mina is the one who made the house a home."

Joan agrees: "Mina gave Nelson the ability to do what he wanted to do. Everybody looks at how Nelson did this and Nelson built that. But it was because Mina was there to support him. You don't want to diminish Nelson at all in the respect that he has quite a story, no doubt about it. Mina is definitely a big part of that story. Mina gave him the ability and the hope. She had made her vows, and she kept them. What her husband wanted to do, that's what she supported.

"People would arrive there, and Mina never complained. She concocted something to feed them. People were always welcome in her house. Mina was very straightforward, but at the same time, she was non-judgmental. Nelson would tend to have his 'this is on the acceptable side and this isn't' sort of thing, but Mina wasn't that way at all. Brian could go there drunk, and he was always welcome at Mina's table. She didn't judge him about those things. She took it like he was just a boy who needed somebody who was stable.

"You think of a little humble fishmonger woman of Scotland, that was Mina—that build and size and Scottish accent. But there will be lots of stars in her crown."

Nelson II remembers his mother as "well-educated, brilliant, with the patience of Job, kind, gentle. She always wanted to live in the city, but she submitted to her husband's plans and made the best of it." Madge was closer to her mother than to her father. "My mother was a very special person. No, she didn't have it easy, but she stuck it out." Bill Irving observed that Madge and her mom got along very well.

Bill also saw another side of Mina. "Mina was sweet. But she was a character. Nelson may have been quite a traditional husband, the head of the household. But Mina had her game plan in mind. She was the design supervisor in the house and ensured that all the bedrooms and dormitories were usable and well-stocked. That was her role: the layout of the house, making sure it was what it was meant to be: a place of relationship-building and comfort. She was a full partner.

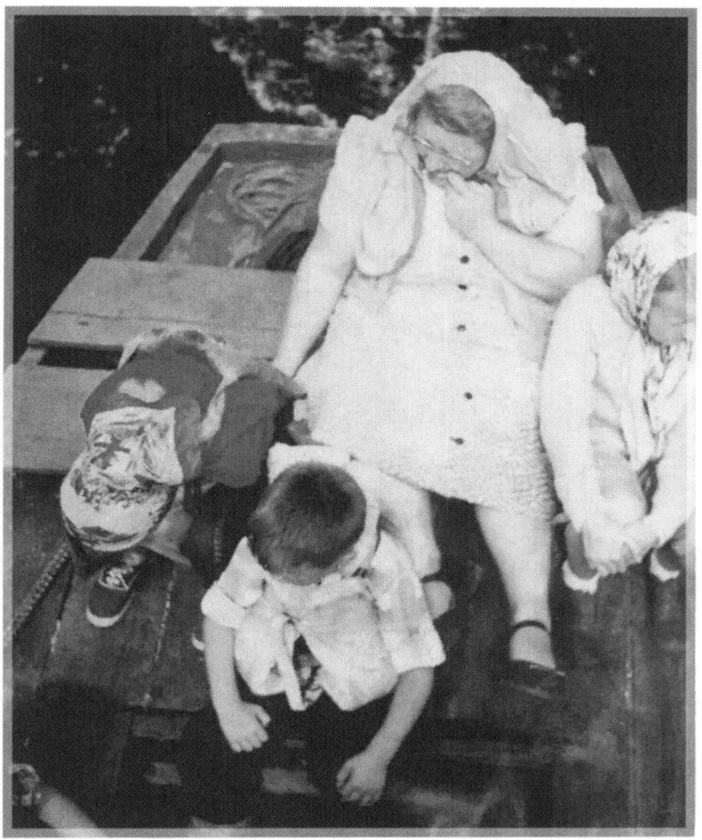

Mina & kids in the skiff, presumably towed by *Raven*, plus a boat behind them.

"Mina often travelled with Nelson to town, and she enjoyed that as a relief from the isolation. It's nice to get out. She certainly enjoyed folks coming in and cooking in her little kitchen. Mina really appreciated when the young ladies would come and stand alongside and help her out. She was quite happy with anything that occurred, whether it was haywire campfire songs or quartet singing. She just enjoyed people that way. And even in her weakening condition, as she got older, she was still up for travel and campfires on the beach. She was always fussing and cleaning up.

"I can't think of any incidents that weren't remarkably appealing in knowing her. While Nelson had some very quiet times and some struggles with different types of relationships, I can't recall that with Mina. She was very caring and affectionate towards Nelson, knowing his ailments and his history. Mina loved fun and wasn't belligerent or obnoxious at all; she loved a good time. She would ask interesting questions, the kind of questions that probe a bit: Why do you believe that?"

Ruth remembers Mina as "very hospitable. She loved to have visitors and cook for them. Living on Copper Island required much fortitude, but Mina was very resourceful. She always had plenty of food on hand in case people came by. Nelson enjoyed the isolation, but Mina loved to have people come. She also loved to play the pump organ and sing the Gospel songs."

Debbie has a story about Lynn, who was a bit of a rebellious teenager. She went to Camp Ross and then to Copper Island. "I think she used some bad language, and Mina took her into the kitchen and sat on her." Joan remembers this too: "And she had a word or two with her. Mina let Lynn know that her behaviour was unacceptable and wouldn't happen again. It quite shook Lynn, who was used to bullying her way around. But nobody imagined that Mina was that tough and strong. Dear Mina, she was this little wee bit of a thing, but physically, mentally and spiritually, she was a power block." Roy added, "She was strong as an ox!"

Debbie remembers, "Mina was plump, and she had a Scottish accent that was so sweet. And no teeth. She wasn't vocal like Nelson was, though. She was the sweet wife in the background who was happy to be there." Joan added, "Her love for the Lord was where she got her enduring strength, after having lived out there in the boonies with no opportunity for fellowship, or other ladies to talk to, and all the hard work she had to do."

Bill Irving remembers that Mina's cooking was very much from the British Isles. "Straightforward, even bland, but certainly healthy. She never wasted anything. I think the carrot peelings often went into a pot for her broth for the next day, which she cooked on the old stove in the living room. There was always the smell of cooking in the house, always something going on."

Mina wrote to her granddaughter, "Please tell mommy that Grandma wants her to buy three small haggis for her this week. Please! Tell her to put them in the fridge, and if your Daddy says, 'Oh no! Not in my fridge, that stuff,' and turns up his nose, you Dear, turn up your nose and say, 'Huh! They're much better tasting than slimy "frog's legs," eh.' If those two abuse you at any time, you tell them, 'I'll send my Grandma after you.' They'll be real scared." Since Leona was one year old when Mina wrote this, it was clearly more of a message to Leona's parents.

Rich told me, "Mina was elegant and cultured in the way she spoke. We always talked about faith and the church and the Bible, and she felt she could trust me. One day, she opened a trunk, and in the trunk was a beautiful set of china. Lace doilies. All the things she had dreamed she would have in the house, when she met the man of her dreams, all still in the trunk."

Several tidbits about Mina appeared in the local news: At a wedding, "out-of-town guests were… Mrs. Dunkin, Madge and Nelson of Kildonan…"[1] About a trip on MV *Lady Rose*, "There was another stop in the lovely little Clifton Bay where

Mrs. Nelson Dunkin joined her husband and her baggage was lowered to the family float."[2] And, "The bride's bouquet was taken to Copper Island for the family's friend Mina Dunkin, who was unable to attend the wedding due to ill health."[3]

Mina was failing in health. Nelson II recalls, "Mom never was that well. She was sick for a long time, with stomach problems. Eventually, it was diagnosed as stomach cancer. I was not living there at that time. She was admitted to the old St. Joseph's Hospital in Victoria, when it used to be down by the harbour. They operated on her, and the cancer went into remission for a time."

Mina in the hospital in Victoria with Nelson and daughter-in-law.

Debbie took Nelson to visit Mina at the hospital in Victoria. She remembers the day when Mina was released. "Nelson and I took her to an ice cream place because she really wanted ice cream. But she was so sick, she couldn't eat it. Those things were hard for Nelson to watch."

And the cancer came back. Rich said, "The prognosis was that Mina was going to die. Nelson built this beautiful room, finished in drywall, off the main living room. There were paintings on the wall, very tastefully done, and he built it for her to die in. Maybe he was trying to give her the luxury that—in their rustic lives together—he was never able to offer her. It was very touching to visit her a couple of times and talk with her about her heart and about God. Mina was deep, and she was quality. Remarkable woman."

Heather remembers her Christian school group being there when Mina was in bed in the room off the living area. "We all sang Scripture songs to her in her room, and Nelson was very caring of her." Nelson II continues, "At that time, my father relented, and they rented a suite in Port Alberni and stayed there, though he went back and forth to the island. She didn't live much longer, though."

Brian was among those who looked after the island property during Mina's last days. "It was the Christmas of 1976, and Mina was in the hospital in Port Alberni. I went and stayed at their home on Copper Island so Nelson could be with Mina. Well, he wouldn't have left otherwise. I was involved at the Bible camp at Pachena Bay but was able to be free over the Christmas holidays. My grandmother was in Bamfield, so at Christmas, I came down and got her and took her back to the island for Christmas. We talked about Christ's birth, and I ended up leading my grandmother to the Lord."

Leona remembers, "I would have been six years old when Grandma passed away. So she was quite young. The other day, my mom said to me, 'You are my mother,' meaning I was exactly like my grandmother. I'm always the one hosting everything; I'm always the gatherer of people. That's just in my nature. My mom is not like that at all, but she says that's the way her mother was, right up until she passed. It's special to realize you have the qualities of your grandmother. I can feel that I was loved by her without having spent much time with her. She's in my life to this day."

Did her illness shake Mina's faith in God? For a time, yes. Joan related, "When she got sick and was in the hospital in Port Alberni, some holiness Christians convinced her that the reason she got cancer was because she had sin in her life. And poor Mina…" Roy stepped in, "She came to me for help on that one. And I said, 'No, Mina, your relationship with the Lord couldn't be any closer or righter. No, you're getting bad information. You can just drop that.' She needed to hear it from somebody else." Ben remembers that when Mina was brought to the church for prayer, "it was one of my saddest occasions when she wasn't healed."

Sad, but not without hope. Nor without a legacy. Mina's resilient faith, which had carried her across the sea and across a continent, stayed with her to the end. Not many people are so well remembered and so well loved, 46 years after their passing. Her gentle influence remains in many hearts to this day, and as for the stars in her crown—they are forever.

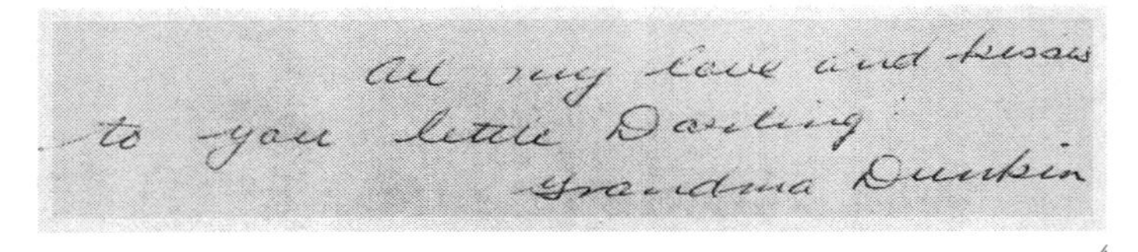
All my love and kisses
to you little Darling
Grandma Dunkin

[4]

---

[1] "Irene Finnie Bride Of William John Andrews," *Nanaimo Daily News*, September 14, 1961

[2] "Day Trip to Bamfield," *Alberni Valley Times*, August 15, 1968

[3] "Lori Mathews weds Cranbrook, B.C. man," *Alberni Valley Times*, July 17, 1977

[4] Letter from Mina to Leona Dolling, age one, January 18, 1972

24. Mina's Passing

PORT ALBERNI - Copper Island resident buried locally

The love and respect for the late Mina Dunkin was shown by the hundreds of friends from near and far who gathered at the Chapel of Memories Nov. 7th in Port Alberni.

Mrs. Dunkin was born on March 9, 1914 on the Isle of Lewis and came as a war-bride to Canada. Her first home was at Port Albion where her husband, Nelson, was employed with Nootka Bamfield Packing Co. Later on the couple moved to Kildonan for a time. [1]

DUNKIN: On Wednesday, November 2, 1977, Mina Dunkin, late of Bamfield, B.C., passed away in the West General Hospital at the age of 63 years.

Survived by her husband, Nelson; one son, Nelson Jr. of Nanaimo, B.C.; one daughter, Mrs. A. (Madge) Vallee of Port Alberni; four grandchildren; and three sisters in Britain.

The Funeral Service will be conducted from the Chapel of Memories on Monday, November 7, 1977 at 1:00 p.m. with Rev. P.E. Wills officiating. Interment to follow in Greenwood Cemetery.

Flowers most gratefully declined. Donations to the West Coast General Hospital Ladies Aux. would be appreciated by the family.

Funeral arrangements in care of The Chapel of Memories Funeral Directors, Port Alberni. (Garry Coderre, Mel Boire and Gerald Hagel Funeral Directors).

For the past 17 years Mina and her husband have lived on Copper Island. Nestled in a silent cove and the rustic setting of the woods, the unique home, constructed by Nelson, holds many precious memories. Visitors who called in while on boating ventures have found the humble dwelling of the Dunkins was a place of tranquility, humour and friendship.

Mina, a lady given to hospitality, demonstrated not only love but a keen interest in each individual for their highest good. Young and old, rich and poor always received encouragement in their visits with Mina.

Rev. Harold Peters of Union Bay recalled his first visit as a Shantymen missionary to Port Albion. The warm welcome he received at the Dunkin home developed into a lasting friendship.

Pastor C. Miller of Chemainus also paid tribute to the pleasure and blessing it has been over the years to have had friendship and love in knowing the Dunkin family.

Danny McLean testified to the influence Mina had on his life. In his memory of her he sang "Just a Closer Walker With Thee" accompanying himself on the guitar.

Rev. Percy Wills of Victoria, pioneer missionary of the Shantymen's Christian Association, presented in his message the real value of the resurrection of Jesus Christ. Faith in this living Christ was demonstrated to all by Mina in her radiance and strength in the hours of sickness and distress.

To mourn her absence are her husband Nelson; daughter Madge (Vallee); Nelson Jr., Nanaimo; grandchildren, two sisters in England and one sister in Scotland. Burial was at the Greenwood Cemetery, Port Alberni.

Many old acquaintances and friendships were renewed at the afternoon luncheon and coffee hour, provided by the ladies in the Arrowsmith Baptist Church.[2]

Card of Thanks

Words cannot express our sincere gratitude to our many friends & relatives who have sent cards, flowers and many other expressions of love, concern and sympathy during the long illness & passing of our beloved wife & mother, Mina Dunkin. Special thanks to Dr's Waite & Donald of Victoria. Also to Dr. & Mrs. Carter and the wonderful nursing staff of the West Coast General for their interest, love and kindness. Thank-you to all the ministers who offered so much comfort and to the ladies who prepared the refreshments after the service. God Bless you all.

- Nelson Dunkin, Nelson & Lavern Dunkin and Andy & Madge Vallee.[3]

The Nelson I met a year after Mina's passing was, in many ways, unlike the person he was when she was still with him. Some of the changes were positive and some were less so. Nelson II recalls that his father "didn't take care of himself at all. He went back to the island and completely immersed himself in doing what he was doing and building." Ben saw this too: "I never heard Nelson refer to Mina very much after she passed. And he did not look after himself as well as he did when she was around. There was dirt in his creases, and he didn't eat as regularly after that. But he never complained—at least that I heard. He was a man at peace with his God and at peace with himself."

But he never got over the loss of Mina. Madge and Leona, among others, remember that "he marked on his calendar, every day, how long it had been since mom passed away. He did this for 21 years. Yeah, he definitely missed her." More loss was to come. A few years later, Nelson II's daughter Arlene passed away at the young age of 12. "He absolutely idolized her," Nelson II remembers. "This loss affected him even more, I think. And he took in her little dog, Snuggles, to live with him."

Bill Irving observed, "I think when Mina passed away, it was a huge adjustment for him in his sense of purpose. You could see him really struggling for some time, and then as people picked up their visits and there were opportunities for folks to come down and share with him in different forms, he re-emerged as an old, very socially aware, active Nelson Dunkin." Nelson II agrees, "I think he became more people-oriented after mom was gone because, up until then, he let mom do that. Now he had to be the social one. So he became more involved with people."

This was the Nelson I came to know: quiet, gentle, humble and ever-welcoming. In a way, knowing Nelson was like knowing Mina too, because there was much of her in him. Nelson could have receded back into his default reclusive nature; instead, Mina's passing made him blossom into a personality that honoured and preserved her presence on Copper Island.

---

[1] "Obituaries," *Alberni Valley Times*, November 4, 1977

[2] "Copper Island resident buried locally," *Alberni Valley Times*, November 15, 1977

[3] "Card of Thanks," *Alberni Valley Times*, November 17, 1977

25. Coastal Missions

The connection between *Messenger III*, Camp Ross and Copper Island has already been noted. Nelson and Mina were the kind of people whom the Shantymen were called to serve, but the Dunkins were also a source of help, encouragement and prayer in turn. As Nelson considered how "God's 105 acres" should be used, this connection became all the more important to him.

The staff and trainees at Camp Ross extended the reach of their ministry beyond summer camps, first to the families of the nearby village of Bamfield and then to the isolated individuals and villages up the Coast. Between 1976 and 1978, five missionary voyages were made by the Camp Ross crew on Brian's fishboat, the *Kolberg*, in the off-season. They reached out to many areas of the BC Coast, Haida Gwaii and southeastern Alaska.

The staff at Pachena in the summer of 1978. Many of these would become the crew of the new Coastal Missions a few years later. Author: second from right.

By late 1979, as the group outgrew the organization's mandate and the needs on the Coast were so evident, the formation of a new mission was discussed. The response of the Coastal community was enthusiastic, and the missionaries scrambled to set up an office and manage the unsolicited gifts that were pouring in to create a new society. Before this happened, a large donation made it possible to procure a 48-foot former government vessel, *D.M. MacKay*, which was re-registered in 1980 as MV *Coastal Messenger*.

At the time, Nelson approached Roy with an idea: Why not make Copper Island the base of operations for the new missionary society? As much as he appreciated the offer, Roy knew in his spirit that this wasn't practical. Nelson had offered before to have Camp Ross run their camps on the island, but the cedar-clad buildings were such a liability, and access was difficult. Roy had to decline Nelson's offer. "Not that I ever had any trouble with Nelson—I greatly enjoyed him. But he also had a mind of his own. So it was awkward when he got upset with us because we didn't use his property. But it just wouldn't have worked."

The crew that had called Pachena home for many years relocated to Chemainus, and they prepared their new ship in nearby Ladysmith Harbour. "On May 7, 1980, Coastal Missions Society became a reality, starting with eight full-time workers—Uncle George, Brian, Uncle Roy, Ron, Petunia, Anne, Debbie, and Gloria."[1] A year later, Tom—who had served at Pachena in 1977—returned and joined the mission. "These were not a collection of look-alikes but a group of very different individuals who, when working together, became a colourful and amazingly skilled team."[2]

For the next 42 years, that largely unchanged group plied the West Coast from Washington State, up both coasts of Vancouver Island, to Haida Gwaii and the Alaskan Panhandle. They brought companionship, cookies, encouragement, prayer and the Gospel to hundreds of coastal people, including Nelson. In need of a better vessel, the crew built their own 52-foot steel-hull *Coastal Messenger*, completing it—in just three years—in 1998. "The survey for the ship's insurance noted 'no deficiencies.' Such a thing is almost unheard of in marine surveys."[3] *Coastal Messenger* usually showed up at Copper Island at the end of April or early May, on its way up the Coast to Haida Gwaii.

Tom remembers his first visit to Nelson on the old *Coastal Messenger*. As they arrived, they discovered that all the meat that was being kept cool in the bilge had spoiled. Tom recalls, "Everything was hectic over the rotten meat, and I was new and couldn't help out with anything. But I liked Nelson instantly. He was the first real person I met who was coastal. Having come from California, to meet Nelson—well, you knew he wasn't somebody from the city, and what you saw was who he was. No pretence."

The new boat was a large step up from the first wooden vessel, Brian remembers. "The generator in the old boat was a one-cylinder diesel, and all you could do was charge batteries with it. When we got a freezer, we had no way to supply power to it. Nelson wanted to give us his generator, and we said, No, we'll get it working for you. You keep it. But no way did he want to keep that generator. It had a glow plug, which takes electricity. And no hand crank or anything; so to him, it was just a headache." So Nelson's generator was installed in the *Coastal Messenger*. "Nelson wasn't into electricity. He went back to wood fires and kerosene lamps."[4]

Brian regularly gave Nelson a haircut on those visits. He would cut a hole in a garbage bag, pull it down over Nelson's head and do the job. Nelson always looked forward to that.

Messenger III—owned by Bill Noon—meets *Coastal Messenger* near Valdes Island.[5]

This past year, the entire Coastal Missions crew retired, if missionaries can ever be said to retire. The ministry and all the society's assets were passed on to new leadership, headed by Catherine Buschhaus, under the umbrella of the International Messengers Canada Society, and with the new name, Coastal Light Ministry.[6] A recent post acknowledges the heritage they bear: "We visit community members who live on the remote coast the same way we visit friends in town. Our visits build on a long history of mission boats, such as the *Messenger III*, and previous crews on the *Coastal Messenger* who served as Coastal Missions from 1980 until retirement in 2022."[7]

My wife and I went to one of the last Open House gatherings at Coastal Mission's waterfront headquarters near Chemainus, purchased in 1985. The original crew was there, most of whom I have known since my summer at Camp Ross in 1978. Every year since that summer, I received a birthday card—and later, one for Sarah, plus an anniversary card—designed by Debbie and usually signed by the whole crew. I have always admired their faith, their selfless dedication and their determination to never ask for money. I have witnessed God's faithful provision for their needs and their consistent friendship despite our infrequent contact.

At that Open House—and after having received from Debbie the cover photo that captures Nelson's twinkle so well—I decided to start a project that I had often considered and should have launched years ago: the writing of the story of Nelson Dunkin. Thank you, Coastal Missions team, for your godly influence and inspiration!

Then came the Coastal Messenger *with Brian & Anne and Tom & Debbie, and you can imagine the time we had and how i fared with the good cooking.*[8]

---

[1] "Coastal Missions: Our Story," See coastalmissions.ca/story.html for the full Coastal Missions story.

[2] Ibid.

[3] Ibid.

[4] Adele Wickett, "Remembering Nelson Dunkin," *Island Christian Info*, June 1998

[5] Both of the above photos are by Tom Maxie

[6] Coastal Light Ministry, coastallight.ca

[7] facebook.com/coastallightvesselministry

[8] Letter to Pat Rafuse, May 12, 1982

Don and Patty Cameron and family aboard MV *Nelson E Dunkin*, which they crafted from the hull of Nelson's boat, *Hera*, that once lay at the bottom of his bay.

26. The Honeymoon Cottage

Don and Patty Cameron characterized a new generation of coastal people. Boatbuilders, fisher folk, homesteaders, dreamers—whatever skills they didn't already have, they were willing to learn, whether the hard way or through the example and tutelage of the older generations. Don and Patty were involved with Esperanza Ministries up the Coast. In the fall of 1979, Earl Johnson asked them to pay a visit to Copper Island, where they would find the hull of a boat that Earl was interested in salvaging. Nelson had been keeping it afloat with a small windmill pump he had made, Peter remembers.

Don and Patty had a troller, and they made their way by water to Barkley Sound. "I remember coming around Clifton Point and suddenly seeing this big old house," Patty told me. "And it looked like—wow, this is pretty rustic and neat. And Nelson, of course, sees us coming, and he's walking along his boomsticks, coming out to meet us. We introduced ourselves, got chatting and told him why we were there."

Nelson had talked with Earl about the boat, a 34-foot double-ended troller named *Hera*, likely one of the boats confiscated from Japanese fishermen during the war. By now, the hull was at the bottom of Nelson's bay. Don and Nelson rigged it up with float logs at each low tide for the next few days, gradually raising the hull up out of the water. "We stayed for several days," Patty recalls, "and this speedboat came up, and we thought, Oh, there's visitors here. It was Earl, and he came by to say, 'Oh, no worries about that boat now because I found another one that's better.'

"We were working at Esperanza that summer and Don's role was to transport people and supplies to Camp Ferrier. Earl said if we did that, instead of payment, he would give us the hull of *Hera*. That was pretty exciting because we were looking for another boat to rebuild since the one we had was old and rotten. This one was at the bottom of the ocean but was apparently better than the one we had. I helped Don rebuild the boat, and it took five years to do it.

"We saw that Nelson had lost a lot of weight, and he was still grieving for his wife, three years gone. And we thought, well, why don't we stay and fatten him up and make sure he's okay? So that's what we did. Nelson loved having us there, and we loved being there. Copper Island is one of those places that is a special portal right from heaven. Staying there grew on us. You pull up at the dock, and it's a whole new world.

"We had recently gotten married, and Nelson really set the course for us. It was a beautiful rhythm to live by—always working on projects and having lots of conversation. It was a beautiful season in our lives, being more unplugged, getting strengthened." Nelson had built a small cabin with a stove and kitchen, and with the arrival of Don and Patty, he designated it the "Honeymoon Cottage" with a carved sign above the door. "He knew Don was Scottish, so he put a Scottish flower [the Scottish Thistle, the national flower] by Don's name, and he carved a shamrock by my name because I'm Irish. He had ideas, and everything had a meaning."

Don and Patty found that Nelson lived a simple life and was happy. "He never lived by the clock but was really in touch with wind, waves, tide, sunrise and sunset," Patty told me.

"Every morning he would read *The Daily Bread*, Don would pray, and Nelson would respond, 'Praise the Lord!' He always said to us, if he was sitting too long for coffee or whatever, 'Idleness is the devil's workshop.' He liked to keep busy, get up and go back to work. But he never worked on Sundays. He still did carvings on Sunday, but not heavy work.

"On Sunday mornings, I always played the three hymns I knew how to play on the old organ. Nelson would dress up on Sunday. He'd wear his Sunday shirt and Sunday pants and suspenders, and then the rest of the week... well, yeah. But we had our little service. We sang a few hymns and he read *The Daily Bread* and Don prayed, so the only difference was that we got to sing on Sundays.

"One night we were sleeping, and early in the morning we heard a sudden big boom, and the bed shook. And we thought, wow, that was a big explosion. Where did that come from? I thought maybe it was an earthquake. We couldn't figure it out. We didn't even know what it was until we got off the island. We went into town or something, and somebody said, 'Oh, Mount St. Helens blew up.' And it was at exactly the same time as the explosion we heard. Man, did we feel it at Nelson's."

Patty enjoyed the rhythm of life on Copper Island. "We got up when the sun got up, and we had our own breakfast in the Honeymoon Cottage. Then we would go and visit Nelson. I'd often make lunch for all of us, and I always made supper for everybody at the big house. Don was usually rebuilding something or helping Nelson, which depended on the weather. I would pick berries or I'd figure out what would make a good supper, which meant rounding up something first.

"The guys worked most of the time, unless there was a day when we were going into Bamfield, and we didn't go very often. But when we did, that's when we got Nelson's mail and our own. If anybody sent us anything, we'd ask Mary. And make any phone calls, like with my family in Saskatchewan, letting everybody know how we were doing because there was no communication. I remember having Christmas on Copper Island. We made homemade decorations for the tree, like construction paper chains. One year we made popcorn decorations for the tree, and Nook tried to eat them."

Don and Patty beachcombed all the logs they needed to rebuild *Hera*, milling up some of them. When they were ready to rebuild the boat, after a couple of years of staying with Nelson, they towed everything back to Tofino. "It was quite a journey," Patty recalls. 'We ended up at the Fourth Street Dock in Tofino and called Gibson's to take the hull out with a crane. The Harbour Master wanted to rent us an old whale museum, and so that's where we rebuilt the boat. It took five years, and in the meantime, kids started coming along. They began helping, too, as time went on." When Don and Patty were ready to launch the boat, they named it MV *Nelson E Dunkin*. Don fished the boat as a troller for several years, and they often returned to Copper Island to visit Nelson.

The *Nelson E Dunkin* was eventually sold to the Shaw family, who fished it for a few years. "Then he had it up for sale, and a fellow was walking the dock one day who wanted to buy a fishboat," Patty explained. He needed a traditional boat for the San Francisco display at Universal Studios in Osaka, Japan. "So he purchased our old boat because it was cute and the shape was really neat." The *Nelson E Dunkin* was placed on a container ship and sent over the Pacific to Universal Studios, where it still resides at Lombard's Landing.[1]

Christmas this time was not like when you folks were here. Just the 3 of us: Snuggles, Hawkshaw and me. Mary did come the day after Christmas and was in such a hurry to go again that there was no visiting. Sunday night before New Year, Al & Patricia came and left again on New Year's Day. It was sure good of them to come all that distance to see this lonely old hermit...[2]

Don and Patty with baby Amy on *Lady Rose*, Christmas 1983
Pencil drawing by Rick Charles, rickcharles81@gmail.com.

---

[1] Search "Lombard's Landing" on Google images and look for a green-hulled boat.

[2] Letter to Don and Patty Cameron, January 3, probably 1981

Nelson and Don splitting cedar boards off a big piece of cedar, with old trading shed in background. Pencil drawing by Rick Charles, 1983. rickcharles81@gmail.com

Notice that Nelson is also walking past the shed with his familiar stoop.

27. The Weather

Nelson's day-to-day life on Copper Island was largely governed by the weather. As one of the wettest places in Canada, Barkley Sound is a different creature on calm sunny days than when a bitter north wind roars into the cove or the Pineapple Express delivers heavy tropical rains. I have seen the rain sluice sideways off the roof there, and mornings so still and sweet it is difficult to set down the tea mug and begin the business of the day. Nelson and Mina had to learn (sometimes the hard way) to live by tide and weather, and not hold too tightly to their hopeful plans. Here are a few weather anecdotes from their letters:

SUNRISE COVE - Copper Island 1987
(Sun on vacation some wheres)

1

How are you standing this winter, weather? It sure slows me down like "molasses in January".

2

i hope this fine weather is agreeing with you both. With all the rainy days we had i just felt like going to the tall timber and a big hollow cedar and saying, "Move over Bruin i'm coming in too."

3

Sunday now and raining: so thankful am i that i got the canoe spliced out while it was dry.[4]

Wednesday night, and blowing a real humdinger southeaster. Wet all day yesterday but last night i said the day would be sunnier after the southeaster and so it was for almost all the day.[5]

Well, this afternoon the sun came out beautifully.[6]

My friends have just left for Port Alberni as the weather is so cold and wet no pleasure in staying... Who can understand the tricks of the weather—now that my friends have left the sun comes out and it looks like summer. Tuesday evening now and the sun is on the other side of the channel. Feeling better today so just staying in and waiting to see what is for me. i would like to do for others but it is all in the will of God what to do.[7]

≈≈≈ This Thursday morning SUNRISE COVE is up to it's Name. ≈≈ [8]

With this fine summer weather i have managed (in spite of the swarms of mosquitoes) to get considerable done on the canoe.[9]

i am right interested in your shop work and the [Noah's] ark you're making—looks like we will need an ark here if the rain keeps up.[10]

Up @ 5:30 this Friday morning and the sun was blazing hot so i expected another hot day as yesterday—then all of a sudden the sun was closed out and cold took over like winter—much like the evil in the world taking over. If Jesus tarry his coming i can see the time could come when it will be a crime to stand for righteousness and to condemn sin. Be as it may i must get out and on with the work. =Tick-Tok, Tick-Tok, Time has passed on the clock= So the sun did come out for a while this afternoon.[11]

Today is a north wind day but dad is going to Bamfield anyway. We are having very bad weather especially the last two weeks. Quite a bit of snow at times. That is why I didn't venture in last weekend.[12]

27. The Weather

Rain without wind today and yesterday was calm without rain so Mary came after some days of southeaster winds and heavy rain.[13]

Sunday night and darkness has crept over land and sea—the bitter north wind still is snarling.

When in winter it snows
And the cold wind blows
Keen to freeze my nose
Sure i'll be blessing that day
i can be hasting away
To where the shamrocks tickle my toes.[14]

---

[1] Letter to Jim and Sarah Badke, April 1987

[2] Letter to Ron Pollock, undated

[3] Letter to Ron Pollock, undated

[4] Letter to Jim and Sarah Badke, November 1, 1986

[5] Letter to Pat Rafuse, December 12, 1982

[6] Letter to Jim and Sarah Badke, February 14, 1988

[7] Letter to Jim and Sarah Badke, May 2, 1988

[8] Letter to Jim and Sarah Badke, May 2, 1988

[9] Letter to Jim and Sarah Badke, August 16, 1987

[10] Letter to Jim and Sarah Badke, April 1987

[11] Letter to Pat Rafuse, May 12, 1982

[12] Letter from Mina to her granddaughter, Leona Dolling, January 17, 1972

[13] Letter to Ron Pollock, October 31, 1982

[14] Letter to Pat Rafuse, undated

Nelson eyes a chocolate cake that a visitor has brought him.

28. Conversations and Opinions

One of the pleasures of a stay on Copper Island was that when the day's work was done, there was little to do but talk with one another. Conversation is a lost art in a society in which we are so glued to our screens, large and small. Nelson loved a good conversation. Most often, he was an observer who contributed infrequently—and then often profoundly. Nelson had his opinions, and among the right people, he wouldn't hesitate to share them, but more often he smiled and nodded at what must have seemed to him sheer nonsense.

"We had a number of interesting chats about his religious beliefs, and some of them were interesting," Bill Irving remembers. "He listened to some conservative Christian broadcasts, and you could see a tension in his mind about Christians being isolated from the world and yet being part of it. He was considered a hermit, but he was very much a social person. I think that was a nice combination." Rich adds, "I say with kindness that he was a simple man. The world needs more men like him. But he was very simple. He was really black and white, and he hated the colour grey."

"Going out there, in the evenings you read your Bible and stuff like that," Wendall recalled. "It was peaceful and quiet, very different; it was getting away from everything. You had your lamp or whatever, and you went to bed early. It wasn't like, well, we can watch TV. There was nothing like that on the island. It was all communicating." Heather said Nelson loved having people visit him. "He was quietly pleased we were there. We just sat at the table and talked with him."

Bill Priest remembers that Nelson was not very conversant. "He was mostly a good listener. It seemed like any topic was of value to him. When I was with one or two others, we would mostly carry the conversation. I don't recall him sharing much." But Nelson loved having people at his table. He interrupted a letter to write, "Well now, Praise the Lord! Bill Priest has just pulled in!! Bill stayed for a good while and we had much talk. i am sure pleased that he came in."[1]

Portrait by Jay Bradley, Port Alberni

Nelson and Snuggles by unknown artist

Nelson was more likely to express his opinion in his letters. Here is a selection that samples his values, principles and beliefs, with enough variety to encourage or offend any reader:

> *Jesus can heal. That expression i do not doubt so much as saying, "Water is wet." Naturally, Jesus can heal. He heals all the time: a sorry plight we would all be in if he didn't. Doctors can cut into our innards and sew us up again but if Jesus never healed—we would only be a pile of painful pieces held together by string. Jesus will heal whom he will (i know because he healed me).*[2]

i was sickened yesterday to read that they are killing babies in Port Alberni. i had assumed that Port Alberni was free from that curse but "no," Satan gets around.[3]

Now concerning the book on witches and spooks [C.S. Lewis' *The Lion, the Witch and the Wardrobe]—i read it from cover to cover and must say i have never read the likes before. It would be unfair of me to make comment on it at this time as i am too woozie-in-the-head to properly think things out to try to understand the reason behind the writing.*[4]

Men of great learning believe (or teach it anyway) that, given enough money, eventually they will know from whence the earth came, whether earth came from numerous grains of sand bonding together or broke off from a much larger mass. Now after they have made up their minds and printed books on the subject, explaining in detail how it all took place, what if some ignorant fool like me should stand up and say, "But Sir Many-Degrees, where did the big hunk come from?" Foolishness.[5]

Question: The human tail-bone: Is it receding or progressing; after millions and millions of years will it be long and useful like the tail on a monkey or will it disappear entirely? Ask your consulting company.[6]

Every day, i find more and more i want to serve Him and be pleasing to His will. i read the newspapers, and the world turns more disgusting to me day by day. Now they are harping on the harlots; laws and more laws; but not a word that there is any sin in what they are doing. Now the Supreme Court of Canada has ruled that stores stay open on Sundays. That is in the interest of Religious Freedom—after all, some Satan-worshipper may want to buy a pack of fags [i.e., cigarettes] *on Sunday.*[7]

It is saddening to see how, in these latter days, the devil is working every foul scheme to break up marriages and homes. Every day we see the traps of Satan drawn tighter.[8]

How we need to pray! People would never guess it but all the night long, whenever i waken, i pray.[9]

It is wonderful how God answers prayer: i was given a large commercial sewing machine (can sew ox hides) so i was praying that i would be shown someone needing the sewing machine you sent up. My son, daughter-in-law and two children 4&5 came for the day. My daughter-in-law had no money to buy a sewing machine and was praying for one so all things work together for good.[10]

This afternoon i went out on the Point to see if any kind of lumber had come ashore—i told God how badly i needed material to work with but every place i went the beach was bare. i was ready to turn back when i was drawn to look over the rocks, not that i had ever found anything there before, and there lay two long 2"X4"'s. How thankful i was/am to God.[11]

---

[1] Letter to Jim and Sarah Badke, October 1987

[2] Letter to Pat Rafuse, February 20, 1983

[3] Letter to Jim and Sarah Badke, February 14, 1988

[4] Letter to Jim & Sarah Badke, May 1988

[5] Letter to Heather Arnott, undated

[6] Ibid

[7] Letter to Pat Rafuse, May 12, 1982

[8] Letter to Leona Dolling, 1997

[9] Letter to Pat Rafuse, December 12, 1982

[10] Letter to Ron Pollock, October 6, 1985

[11] Letter to Ron Pollock, February 21, 1983

29. Wood Carving

Nelson never once gave me one of his carvings. I have seen many pieces of his handiwork, treasured by other people, but to me he always said, "You don't need anything from me; you can carve your own." He considered my work to be "finer" than his, but his unique signature style is well-loved by the many recipients of his carvings and other projects. At least I had the opportunity to take pictures of his works for this book. I am not (very) bitter.

1

R. Bruce Scott, a frequent visitor, termed Nelson's style "folk art," referring to a less sophisticated, culturally-derived craftsmanship.[2] This is a good description of the wood carvings and metal fabrications Nelson carefully and thoughtfully created. When Dave first met Nelson, he "found him in his basement workshop working on one of the carvings he was well-known for. Each one was unique and usually had a message and some of his unique sense of humour." Most of Nelson's creations were very intentional—made for a specific person with a purpose, memory or lesson just for them.

Anyone who had the privilege of visiting Nelson in the original house remembers best a glassed-in wood cabinet with an upper and lower section. The upper part contained a depiction of heaven; the lower part, well… hell.

It was surprisingly graphic for such a gentle soul to have crafted. Heaven depicts music-making angels, a lion and lamb lying down together, and people watching what is happening on Earth below. Plus there are several other figures with some obscure meaning. Somehow, a boat that looks a lot like his old *Raven* ends up there too.

The bottom section was another scenario altogether. Rich remembers "the old drunk running away from a demon. And Nelson explained that drink was all the person ever wanted, and now that's all he's got." People are fighting one another with axes. Ben remembers that "red, flickering flames appeared around the edges. An ugly Satan with a forked tail carrying a trident stalked the place with a look of evil glee on his face."

3

Witches, vultures, demons and evil creatures abound. A man in despair has a snake twisted around him. Jack McLemon called it "a sermon carved in wood." The cabinet was one of Nelson's many projects that were lost in the fire, and only a few photos of it remain. To my knowledge, the only carving to survive the fire was the game based on John Bunyan's *Pilgrim's Progress*, given to the Sadler family, and now lost.

Celebrating happier occasions were two rocking cradles that Nelson fashioned. The latter of the two was made over a period of three months for Madge's daughter Leona.

Madge told me, "It's beautiful. It has everything on it, because 1971 was the year of the BC centennial, so he included a dogwood [BC's provincial tree], and there's the V for Vallee [their last name]. I had the cradle for all my kids."

The cradle has three verses of "Away in a Manger" carved on one side and Luke 18:16 on the other ("Suffer little children to come unto me, and forbid them not: for of such is the kingdom of God"). One end has Psalm 23:1 with a lamb ("The Lord is my Shepherd, I shall not want") and the centennial dates with a dogwood (1871-1971), and the other end has an angel with Matthew 18:10 ("Their angels do always behold the face of my Father which is in heaven").

Madge thinks that her cradle is the best one. It has been used in the nursery and for Christmas pageants at Arrowsmith Baptist Church for several years, but still belongs to the family.

The first cradle was made for his first grandson, Nelson III. This cradle had the Christmas carol "Away in a Manger" carved on the sides, and at one end, the words from Numbers 6:24-26:

> *The Lord bless thee, and keep thee:*
> *The Lord make his face shine upon thee,*
> *and be gracious unto thee:*
> *The Lord lift up his countenance upon thee,*
> *and give thee peace.*

A Jewish menorah (a seven-branched candelabra) is at the top, and a lamb is carved and painted at the bottom. R. Bruce Scott remembered that Nelson had painted "a bagpipe-playing angel at the head—thereby inferring that all, or most, angels are Scottish."[4] Steve remembers this cradle as "a true masterpiece of dedication and love. Nelson's appreciation for his Creator was revealed openly through his love of woodcarving." This cradle was also sadly lost in the fire.

Nelson was often commissioned to create wooden signs for the community; some are still standing. An example is one

Nelson made for the BC Centennial (this must have been a busy year for him) that still stands beside the Bamfield Info Centre and Museum.[5] The sign is an upright 15-foot cedar plank carved with an eagle, the Centennial logo and dates, and two playful black bears. He included the reference Isaiah 40:31, "But they that wait upon the Lord shall renew their strength; they shall mount up with wings as eagles; they shall run and not be weary; and they shall walk, and not faint." An article lists other signs that Nelson made for the community: for Morrison Hall at Camp Ross, R. Bruce Scott's Aguilar House, the United Church and Ostrom's Machine Shop.[6] Ben remembers another sign that "for years stood beside the rough road and helped people find their way to this town straddling both sides of the Bamfield Inlet." Nelson also painted a sign for the Bamfield Red Cross Outpost Hospital.[7]

Nelson salvaged wood that drifted into the bay or washed up on Clifton Point or Pebble Beach. Nelson told Pat that when he needed wood, he prayed it would come ashore. He would always go log-hunting, especially after a good blow, and though God often provided, it wasn't always in the most convenient location. A common spot for drift logs was Pebble Beach, which was up and over a hill from his house. "Nelson said he knew there was an angel on the other end, helping him carry that log when he needed help." Often, visitors or neighbours would bring him wood, especially pieces they thought he would find inspiring: "So good to get your letter and the slab of juniper which is very interesting. If i weren't so busy i would start on it right away—looks like it would be more firm to carve than red cedar."[8]

9

In October of 1986, Nelson wrote, "Peter of Kildonan was to come and saw some Alaska Mill lumber for me. He was unable to come as he had to go to town with a bad tooth, so

Jim & Dodie showed up with all their milling equipment. The deal was to cut a good 5" plank from a 20' log and the remainder into 1" boards. The large plank was for the Dr. McLean Memorial which Esperanza wishes me to carve. Well, the log turned out no good for carving, all full of little knots, but got some good 24" boards for the canoe. i had offered to pay $100 for the sawing, but no go; Dodie has a little two-wheeled Coffee Mill which they wish me to put into running order. it's new but they kind of botched the making of it. Also, Dodie brought a good dinner plus a chocolate cake, so now what can i do but fit the little Mill?"

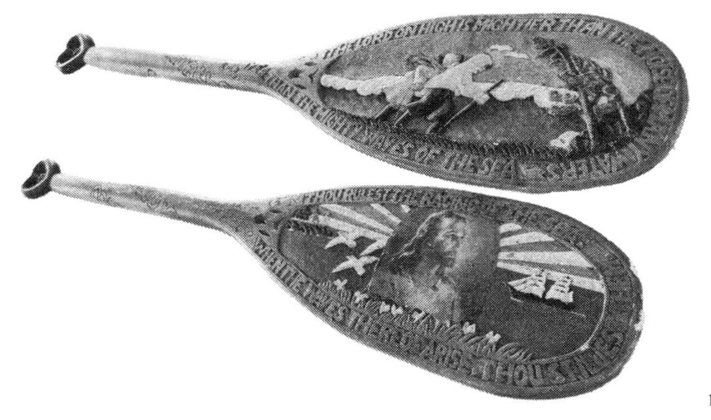

10

Nelson's thoughtfulness often took precedence over the many tasks and projects that normally kept him busy. He continued, "Didn't do much on the canoe this week as i was busy on an Isaiah 40 eagle plaque. There is a poor man in Banfield living all alone, he has just suffered a stroke, has but one eye, and a cataract is taking the sight from that eye. After the stroke, he sent a pile of his engineering books to me. Mary, Lord bless her, says the plaque is very pretty and took it back with her. She said that he told her that he is a Believer."[11]

In a letter to Ron, Nelson wrote, "About the plaque for Heather Arnott—Can you send me a picture of a Red Wing Blackbird as i do not know what they look like and have no description of one?"[12] This was research for a carving that incorporated Heather's favourite bird, flower and Bible verse, which he had asked of her. When completed, Nelson wrote to Heather, "Now concerning the bird: [Ron] asked me to do the plaque for you and i knew right fine that he was wishing to have the joy of giving it to you—so he had the joy of giving it to you and i had the joy of doing it for you and you have the joy of having it. 'It is more blessed to give than to receive,' and in this particular case there was two givings and one receiving. (Please bear with me and my poor English, i have forsaken all hope of education.)"

Another letter describes a project for Mary: "The week before, i built a dog-cottage for Mary. Not for Mary to live in but for one of her four dogs. She now has a dog refuge home; all the forsaken and ill-treated dogs come to her for help. When Mary comes here, Snuggles runs to meet her and gives her a sad story of how abused she is. (In your rummaging Ron, might you come across a dog hide that will fit me? Even a good wolf skin might do.) When i sent the doghouse to Banfield on the *Lady Rose* it was quite the attraction. Of all times, there was a big Gov. boat at the dock, so the *Lady Rose* landed alongside and the freight was loaded ashore by the tackle of the big boat. So the doghouse was swung high for all to see. Had i a representative on the job, i could have taken orders for a number of doghouses. Then i would become a money-hungry Developer. In the Dog World that is."

Ben remembers a mini-wheelbarrow Nelson made for his son David. "Nelson showed his love for Dave by making a small wheelbarrow for him and painting it blue and red. It lasted for years and would likely still be around were it not for it being run over in our driveway. Dave loved Copper Island.

On his first trip to Nelson's house, just a few months old, Dave fell out of bed with a thump that woke up his mom. She promptly tucked him back in bed with her, and Davy didn't even wake up or cry. Dave enjoyed fishing for shiners off the dock more than going out on the boat for big fish."

Bob Miller wrote about "one carving that Nelson said he wished he had never made. It was a game in the form of a life-sized pea pod with five or six peas. All you had to do was get the peas into the pod by shaking the box that they were mounted in. It was surprisingly difficult but all of us managed to master it during our stay. However, Nelson was disappointed at the amount of time that people spent with that 'silly' game."[13]

Nelson also loved to encourage others to create things out of wood by hand. Several people, including myself, tried our hand (and feet) at his foot-powered wood lathe, candlesticks seeming to be the favourite project. Patty mentioned she "made a double candlestick centrepiece for a dining room table that I still use today." A visitor from West Germany did the same.[14] When I mentioned I was looking for a base for a burl table I was constructing, Nelson wasn't satisfied with the available driftwood. Instead, we went and cut down a small tree, and spent the good part of an afternoon digging out the root to use as the base.

And some projects never progressed further than the dreams in Nelson's head: "Please tell me truthfully if you think i am thinking a wrong way to do service to Our Blessed Lord: To make a display of Micah chapter 4, puppets which act and move by machinery which is wound up like a grandfather clock. Could be situated by electric motors and electric magnets but then it would be dependent on power from outside. Winding up a weight—it would be all on its own. i think there is enough of the electric stuff already. It would need to be behind glass and a safety clutch on the wind-up crank—for because of the fallen human nature, it would be great sport (for some idiots) to see just how much abuse the mechanism could stand before breaking."[15]

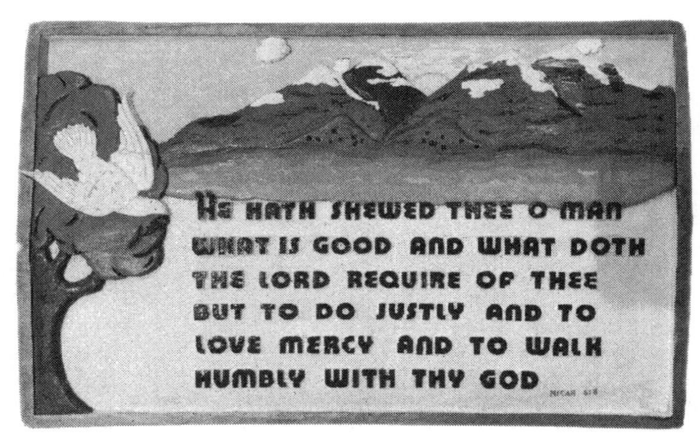

[16]

---

[1] Carving owned by Madge Vallee

[2] R. Bruce Scott, *People of the Southwest Coast of Vancouver Island,* 1974

[3] Photos by Wendall Farrell

[4] R. Bruce Scott, *People of the Southwest Coast of Vancouver Island,* 1974

[5] Photo: Bamfield Community Museum & Archive, bamfieldparks.com/info-center

[6] "Centennial Park Sign," *Alberni Valley Times*, August 7, 1968

[7] Pat Garcia, *Bamfield: Looking Back*, 2010

[8] Letter to Jim & Sarah Badke, April 1987

[9] Carving owned by Pat Rafuse

[10] Carving owned by Dave and Inger Logelin

[11] Letter to Jim & Sarah Badke, October 1986

[12] Letter to Ron Pollock, August 23, 1986

[13] Bob Miller, "Nelson Dunkin," Black-and-White Album, sites.google.com/site/blackandwhitealbum/the-klassens/ann-klassen/nelson-dunkin

[14] Letter to Jim & Sarah Badke, April 1987

[15] Letter to Pat Rafuse, December 12, 1982

[16] Carving given to Heather Arnott, now owned by Brian Burkholder

30. Firewood

During many winter days on the West Coast, the air feels like you are walking through cold water. Nothing short of a hot tub—or close proximity to a roaring woodstove—will warm your bones. Nelson fashioned his own woodstoves for the most part, from coffee can to oil barrel size, and as long as nothing went wrong, they worked remarkably well.

The stove that was neither his design nor in his domain was the wood cook stove. This was Mina's territory, and though she was willing to let qualified guests lend a hand with it, Nelson was not one of them. His role was to keep a good stock of dry fine-split firewood to keep it going. Once Mina was no longer there, Nelson gravitated to the small woodstove in the self-contained suite. The kitchen stove was mostly used when groups came with their own qualified cooks.

I remember well the big woodstove in the living area of the original house. made of two oil barrels stacked on top of one another. Wood burned in the lower barrel, and the top barrel achieved a secondary burn of the smoke and gases released from the wood. It worked essentially like a modern high-efficiency woodstove. The lower barrel had a damper lever that was so airtight, you could hear the change in the burn if you adjusted it the tiniest bit. The big stove and the cook stove shared the same chimney. Mina commonly used the big stove, when it was going, for things like the soup pot or foods that took a while, such as canning fruit and fish.

The tiniest stove Nelson made was for Ye Hermitage, the wee outhouse-sized cabin (which was actually once an outhouse) that Nelson fled to when his house became too crowded. Visitors remember this stove being about the size of a large coffee can (or made from a coffee can), but it managed to warm the wee place just fine when stoked with chunks of fir bark from the beach.

And everyone remembers the fir bark. R. Bruce Scott noted that "although he burned wood for heat and cooking, Nelson did not have any woodpile on hand. When I asked why, he replied that he didn't believe in it; living from day to day was enough for him." Bill Priest watched as "he would go along the shoreline gathering bark for firewood. I found that strange, and then I realized it was a pretty smart collection of firewood because it was light, had no sap in it to creosote up the chimney, and there was an abundance of it floating in from the logging industry debris. He really didn't have a wood pile."

Patty has the same recollection: "Nelson collected bark all the time, and we said, 'Oh Nelson, we'll help you get some firewood for your stove.' He said, 'This is how I do it.' He had this leather bag, and he went out on the beach and got all the bark he needed. That's all he burned. And it burned well when it was wet too—the BTUs in the bark were incredible. He never took logs off the beach for his stoves." Madge made the observation that the salt in the bark continually ate through the metal of the stoves and their chimneys.

And this was the problem. As Nelson would later observe, "Fire is a good slave, but a cruel master." And the house he lived in was entirely built out of cedar kindling.

Must make my nest up and go to bed as it is getting cold and i want to keep the dry wood for the morning fire.[1]

i was out on the beach and gathered bark sufficient for the night. How do you keep your cabin warm?[2]

---

[1] Letter to Jim and Sarah Badke, April 1987

[2] Letter to Pat Rafuse, undated

31. The Fire
by Nelson Dunkin

Copper Island, Easter Sunday 1982. Down in the cabin.

My Beloved Pat:

Yesterday, Don and Al showed up with a 26-foot boat hull which Al bought in Port Alberni. They were 8 hours pushing it with a little outboard motor. Patty came up to Sarita by car. Mary was just leaving there at the time and brought her over. After dark, a boat came from Banfield with my former nephew [Fred Longabaugh] and a couple with a two-year-old little girl. All went well and everyone went to bed.

At 2 o'clock Easter morning, the fire alarm was given. The fire was burning all the roof and upper part so that we had no chance at all of bringing it under control with the one little fire pump so all hands went to work moving things out. My book of poetry is no more but thanks be to God my last piece is on this paper. i had thought it was lost. Your letters i had just put specially in a nice blue box. In the thick of the fire i made it a must to rescue this box and today how happy i am on opening this box to find this sheet in it.

Now was the fire the work of Satan—or was it the work of God to teach me something, that i do not know—at any rate, God allowed it as like the testing of Job. Praise the Lord. All during the fire and now i have a calm feeling of trust in God that He gave me much, and like Job, He can give me more again.

As my son said at Arlene's death, that he was all the more determined to carry on for the Lord, so say i. It must have been the old devil as Copper Island has been greatly used of God. Thousands upon thousands of money's worth of things have been burned. But i am most thankful that my faithful old hammer, saws, and other carpenter tools are saved. i feel just like going at it again. Before this, i was thinking it would be wise to not add more to the old house but to here and there build little cabins hid away in the bushes. Then in case of a fire only one unit would be lost.

For a while, i didn't see Snuggles, but later she came and had me hold her as she was trembling like a leaf, and i am thankful, but poor kitty fared not so well—this morning i saw her burned body laying on the beach.

The organ, an hymn book, Strong's Concordance and my Grandma's Huge Big Bible plus 2 other Bibles are saved. The $500 i had laid away for paying the taxes is saved. Thank God. Also the lighting plant [i.e., generator] and the Sea Gull motor which i bought from Julie Bradford, Tofino.

Patty, Don and Al are going around without shoes.

The Hermitage, Easter Sunday Night

Snuggles and i are here in Ye Hermitage with a cheery fire in the little stove which i built and it is working very well. =FIRE IS A GOOD SLAVE BUT A CRUEL MASTER= They thought to put their little motor on Al's hull and go to the Joe's across at Sarita and on to Tofino and let Wilson Joe tow the hull back to Copper Island but then a mean North wind with rain blew all day. So they will stay another night.

The young couple from Tacoma are very fine people, i wish them to be Christians—they would make good Christians. i keep thinking God must have something for me. My carving tools are not, but am i to think that God wants me to carve no more?

In the heights the angels sing
"Glory to our Lord and King."
Should not we here on earth
Join in with the angels' mirth
And make the wood and meadows ring
As to Christ in joy we sing.

And now Pat Dear
Your turn is here
With your fine touch and skill
The open spaces to infill
With mountains grand and high
And angels singing in the sky
Saints in the meadow many flowers gay
Singing at the break of day.
(printing is no fun
the way this ink wants to run)

[Written on an old grocery order form, probably all the paper he could find]:

For supper we had rice, just rice—the young lady from Tacoma just now brought some rice for Snuggles but Snuggles (with all my coaxing) said, "Do you think me a China dog?"

Ye Hermitage, [Easter] Monday morning

Snuggles and i slept well all night—Snuggles says she wishes Pat were here—i tell her she should not be wishing such things on Pat.

Ye Hermitage, [Easter] Monday night

Rained all last night, Rained all day and still raining.

Don, Patty and Al went to Sarita by Al's hull, then returned by Angie and Eunice's canoe, towing the hull back. Angie and Wilson still in Victoria for dental work. Eunice loaned them boots to wear and sent back a box of food. So that is Christianity in action.

Built almost entirely of cedar, the original house was a tinderbox waiting for a spark.

Young [Bill] Priest (surname) and his wife came from their fish farm in San Mateo Bay. Very nice Christian people. Nelson II and Lavern are due tomorrow and the Priests will come down to visit. He wants to go to Bible college and get on a mission boat.

The Hermitage is warming up and Snuggles is snoozing.

i neglected to thank you for the violet you sent me—Thank you, Pat. That little flower means much to me. At this time i keep thinking of a Scripture, "Behold, i make all things new." And i have faith that God will do just that.

So very sleepy—Good Night.

Tuesday. Very big rush. *Lady Rose* in with Nelson II but he has to go again as he has a funeral tomorrow. Also Fred, and the couple, went to Banfield to go on to Tacoma.

Much people have sent all kinds of food, clothing and things. Praise God.

- Nelson[2]

This is the amazing account penned by Nelson in Ye Hermitage as, close by, his house lay in a smouldering heap in the cold wind and sleet of a chill Easter weekend. It is astounding that his letter begins, not with "My house just burned down," but with an account of who was visiting him at the time. Even more astounding is his attitude toward the great loss of not only the house but all the carvings and mementos of his life on the island with Mina and the children. His "calm feeling of trust in God" during the fire was the result of a lifetime of faith, and it would carry him forward.

Don and Patty Cameron had moved to Tofino but were visiting Nelson on that Easter weekend. Patty was the first to notice something amiss on the Saturday night. Here is the story in her own words:

"We were downstairs visiting with everybody, and Nelson had a woodstove made from 45-gallon drums. It was going pretty hot, and he kept adding wood in there and we thought, wow, it's really burning nice. We were all cozy and we were all visiting. It was starting to get late, so we thought we'd turn in. Don and I were staying in the room upstairs across from the bathroom. Everybody was sound asleep, and I sat up and heard the bathroom door sucking shut, and then opening and sucking shut again. I thought, well, that's odd; I haven't heard that before. Assuming there was nothing to worry about, I laid down, and right away Don shot up.

"He ran to the window and saw orange flashes all over the trees outside. He said, "The house is on fire, honey." It was pitch black, and he was trying to get his pants on but they were mine, and he couldn't get them on. I tried to grab a few things and he said, "I'm running down the hall; I'm telling everybody to get out. Honey, you and the lady and the little girl run all the way out to the very end of the dock and stay there." So it was me and this lady and the two-year-old out on the dock.

"Don got all the men up, and they started throwing water on the fire. The chimney pipe, where it came out through the roof, had only flashing around it, and it had gotten really hot. They splashed a bunch of water on that spot, but just a second later, the whole roof was going. So the men began trying to get stuff out. Nelson was trying to go up the stairs, and Don picked him up, put him over his shoulder, and carried him out. Nelson's hair was all singed on the top of his head. They got everybody out and they grabbed the pump organ and whatever else they could.

"It was windy out at the end of the dock, and the lady and the little girl and I watched the whole house go up in flames. We saw the frame burning and the bathtub falling down from upstairs, and things were exploding. It took four hours and the entire house was down, and there were like six to eight feet of coals just glowing. The fire created such a wind of its own. We could feel the heat from the end of the dock. We couldn't even move or do anything until the guys came to get us and said it was safe, and they took everybody to the Honeymoon Cottage.

"So the little cottage we used when we lived there had everybody in it. We had no blankets, no nothing. Somebody had rescued a box of rice, so I cooked a pot of it. I remember trying to feed everybody. Back when we lived with Nelson, every morning before breakfast, he always read the *Daily Bread* and asked Don to pray. And when Don said Amen, he'd say, "Praise the Lord!" And you know what? That morning, with the house burned down and we're all in the cabin with everybody, Nelson read the *Daily Bread* as if nothing had happened. Then Don prayed and Nelson said, Praise the Lord!" He had lost his whole big house and all his belongings, all his carvings, everything, and he still said the same thing.

"None of the neighbours saw anything in the bay. We were stormed in, and we couldn't get out for two days. And it snowed—it was really cold that Easter. I found a pair of runners of mine in the Honeymoon Cottage, so I put those on. But the guys didn't have any shoes. They found some sacks somewhere and tied them around their feet so they could walk and get some wood and just try to manage. We couldn't even get the word out. We didn't have anything.

"After the second day, Al took the boat we had sold to him—it was a wooden boat we rebuilt—and he was able to get us across to Sarita River, where Angie and Eunice lived, the two native sisters who were really good friends of Nelson. We told them the house had burned down. We had our car not far from there, up the Sarita River, so we hopped in and went to Port Alberni and told Madge about the fire."

- Patty Cameron

Nelson II remembers his father's calm attitude about the fire. "When I heard about it, we loaded up a whole bunch of supplies, and I took *Lady Rose*. When I got there, he was living in the cabin and was as content and happy as can be. He basically said, This is the way it is; I'll just go on. I thought he'd be devastated, but he didn't seem to be. He seemed to think, Well, I'll get up tomorrow morning and start doing it all over again."

Earl commented about the fire, "He lost everything but still had everything. We lose it all and we've lost nothing. Nelson knew that."

Rich sums up the story of the big fire: "The night the house burned down, as they watched the flames, Nelson said with his stammer and his inimitable humour, 'You are my special friends—I wouldn't put on a show like this for just anybody.'"

Nelson's binoculars saved from the fire, now owned by Brian Burkholder

---

[1] Bamfield Community Museum and Archives

[2] Letter to Pat Rafuse, April 11, 1982

32. The New House

Word about the loss of Nelson's house spread quickly throughout Barkley Sound and the West Coast. The first response was to send Nelson an impressive amount of stuff, some of it useful, some less so. "He seemed awed that so much stuff came to him," Nelson II remembers. "We collected all kinds of utensils, food, dishes, blankets—all household stuff. I think several things were given to him that he wasn't sure what to do with. He was actually giving stuff away, which was his habit. He gave stuff away all his life."

The Sadler family, particularly Jim and Gil, along with Bill Irving, were the primary architects and builders of the new house. Jim's son Harold remembers helping out too. "We came down a few days after the house was burned and started building a new addition next door," Bill told me. "We went down probably three or four times and helped in different stages of construction as it was going up. But you're only there for a day, and you do what you can and then you take off. So Nelson kept plugging away at it. And we helped run the paths and water, and the sewer was pretty straightforward. There was always good fishing by the outhouse at the end of the dock."

Patty laughed about some of the things people thought Nelson could use, including a whole box of purses—perhaps for the leather? And a roll of blue carpet. "Nelson looks at the carpet and he says, 'What am I going to do with that? I don't have a sucking machine.' The next time we came back, he had lined the walls with it as insulation. It looked like ocean, so he carved ships and attached them to it; he made it really cute."

Ben recalls, "Nelson was well-loved around Barkley Sound, and many began to plan, offer practical help and give donations to build him a new home. Nelson was not above working on the new home himself. On one visit, I saw him clinging to an upright timber and pounding nails with his other hand as he built another room. He was not afraid of work."

Indeed, he wasn't. In May, he wrote, "This was another sunny day and i got along at the building of a sanitary house down at Honeymoon Cottage. Something needing to be done for years now. i have finished the two-room one out back here. Just waiting for you to come and plant flowers in the two hanging planters which i made for the porch. The name is 'Pollockville.' Ron was here when i started to build and i asked him if he had a name for it and he replied, 'Call it Pollockville.' So Pollockville it is, in 6-inch letters carved into a cedar timber."[1]

During the several months when the new house was being built, Nelson stayed in the Honeymoon Cottage—tight quarters when visitors came, as Sarah and I once did, not knowing until we arrived that the old house had burned down. I remember the cottage as being dark and close. There was also the one-person Ye Hermitage, where on occasion Nelson slept to escape his visitors. Patty remembers that this cabin used to be the double outhouse on the dock. When the kids wouldn't stop fishing there, Nelson decided to pull it up on shore and make it into a one-bed cabin. Patty made the curtains for it.

But despite all the assistance, everything took a long time. At the end of October, Nelson wrote to Ron, "Wynn is a very great help here. Yesterday he nailed the last shake on the side of the new little house. Working all alone, and with the stormy weather and short days, i would be all winter at it. As yet the chimney builders have not arrived—if by the time Wynn returns, if still no brick layers, we will go at it ourselves."[2]

Later, he wrote to Pat, "Wynn Boothroyd is still here and a great help. As yet the house is a ways from being ready to move in. Strange, but everyone in town seems to think that i am in it. There is little carving or woodwork i can do until there is a place to work." And later again, "i am most thankful as the chimney is now up to the roof ready to have a hole cut before it can go higher. So the time draws nearer to having a house and i am wondering if it will not be a lonely house."[3]

The new house was two storeys tall and narrow, a mere fraction of the size of the former house. A dark and narrow staircase took you to the second floor, which was his living area. There was a small stove and kitchen, a nook with a table and bench seats, his bedroom at the other end, and in between, his chair facing a large window with a view of the bay. Beside it rested the old pump organ, for whatever reason, as he didn't play it himself. The lower floor was his shop. It soon filled with tools, equipment and materials for his various projects, which if anything were more numerous and ambitious than before.

A gift from Nelson to his granddaughter Leona in the year of the fire.

---

[1] Letter to Pat Rafuse, May 12, 1982

[2] Letter to Ron Pollock, October 31, 1982

[3] Letter to Pat Rafuse, December 12, 1982

33. Creatures of the Land

Nelson grew to love people, but every creature loved him. The story of Nelson couldn't be told without including the animals, birds, fish and mammals of the sea with which he interacted every day. Patty remembers Nelson's gentle spirit: "He would pick up a piece of bark or wood, and if there was a spider on it, he'd always blow it off first. He would never burn a spider on firewood; he shooed them off. However, he thought he could invent a new glue or something with slugs."

Probably the most memorable animal on Copper Island was Nook, Nelson's massive Newfoundland dog—like a Saint

Bernard, but black. The first time I went to Copper Island, I took my lunch outside and sat on the walkway with my legs dangling. As I went to eat my sandwich, a large, hairy, drooling muzzle intercepted my efforts. Though he didn't try to take my sandwich, Nook was making sure it never reached my mouth either. There was nothing I could do to prevent him from lodging himself between me and my sandwich, except to tear off a chunk and toss it down on the beach. I had to quickly finish my lunch before Nook loped back into position again.

Brian tells me that Nook was named for the bay across from Copper Island, Poett Nook, which is where the dog came from. Poett Nook, in turn, was named in 1861 by Captain Richards of the HMS *Hecate* after an English physician who was visiting from San Francisco and had an interest in copper claims on Copper Island.[1] Nook was often mistaken for a bear by the tourists watching from the railing of *Lady Rose*. Heather remembers that Nook would greet visitors by enthusiastically slobbering on them, nearly pushing them off the haphazard dock.

Nook was a thief, and knew it. Brian told me, "The kids loved catching perch off the float, and they probably caught lots of shiners, too. They would catch a fish and put it on the float. Well, Nook would steal their fish, and he knew he was stealing it. He would get it when they weren't looking, and he'd walk up the beach before he would eat it." One time, Debbie had made Nelson a beautiful chocolate cake and set it down on the dock as they were unloading. "When we weren't looking, Nook ate the whole cake and had it all over his face."

Everyone remembers that Nook was one stinky dog. "You didn't want Nook to brush up against your clothes," Roy recalled. "Nook had a very strong odour. And he was so happy to have people visit, too. He was just as happy as the people were. He would run down to the float, and be all over you and shake. Yeah, he was a typical Newfoundland… No, he wasn't typical. He was his own person. Just like the people he lived with." Nook also had a hair-loss problem, possibly due to his diet of raw fish—when he couldn't find chocolate cake.

Despite his shortcomings, everyone loved Nook. Brian said, "Nook was actually a very smart dog. He would line up with everybody to have a picture taken. And after he heard the click on your camera, he would come for a treat. Somehow, he had learned he was supposed to get a treat after he lined up for a photo." Nook often hung around the kitchen, hoping for a treat. "We would toss it out the door," Patty remembered, "and the dog would go flying over this little treat." Mary was one of Nook's favourite people for the same reason—she always had treats for him.

Nook passed on not long after my first visit to Copper Island, which was a year after Mina had gone home to Jesus. This must have been a very sad period for Nelson, because not long after, his first granddaughter Arlene, whom he loved dearly, passed away. Arlene had a small mixed-breed yellow dog named Snuggles, which Nelson took in as his own. Unlike Nook, Snuggles could curl up on Nelson's nap as he gazed out the window. He grew to love the little dog, who proved to be a faithful companion.

Snuggles was there when we visited Nelson with our son Ben, who was eight months old. Nelson wrote to us afterward, "i haven't told Snuggles that Ben is not easy about her as it would hurt her feelings very badly; she is so very fond of little people and tries to please them. When you come we shall have to get the two of them together in understanding. Sometimes i read parts of letters or articles to Snuggles—when i came to the part about Ben's interest in crabs, Snuggles made a mad dash out the door and away to the beach—soon she returned with the enclosed crab—very happy to have me dry it and send it 'to Ben from Snuggles with love.' She sure likes little people as soon as they learn not to poke their fingers in her eyes."[2]

Author's son Ben with Nelson, and Snuggles looking less than impressed, 1986.

Copper Island was not without its hazards for small dogs—eagles and the occasional cougar among them. Nelson wrote to Pat, "From hence onward Snuggles has solemnly promised that she will never again harass the seagulls. The other day the seagulls were the means of saving her from drowning. Wynn heard the gulls making an uproar, then he listened and heard a splash splash—he ran down the float just in time to pull Snuggles out before she submerged."[3] When Nelson grew ill and had to leave Copper Island, his daughter Madge took the elderly Snuggles to live with her. "We always joked that Snuggles was my inheritance," she chuckled. This dog must have enjoyed a long life. In a letter to Leona in 1994, Nelson wrote, "i am still very sad for Snuggles."[4]

Someone with good intentions brought Nelson another cat to replace the one lost in the fire, and he called it "Hawkshaw," a reference to an investigative detective. Patty described the house Nelson made for the cat: a door cut into the bench that ran under the front window, with "Hawkshaw" printed over the top. Nelson wrote, "Cats, i think, must enjoy their times of sleeping, or should i say catnapping, judging by all the various positions he (that is, Hawkshaw) assumes." But later he added, "A beautiful summer Wednesday morning. But i just don't know what to do about Hawkshaw as it's costing altogether TOO MUCH to keep him."[5]

Copper Island is home to a variety of wild animals. Bears and cougars have been seen swimming the channels between the islands, and both have been sighted on the camp's trail cameras. Deer love grazing on Nelson's property, and I have often seen spotted fawns romping on the grass. Nelson II said that wild animals gravitated toward his father. He was always amazed to see animals come right up to Nelson without fear, like Oscar the river otter, who would let Nelson pet him.

Some animals were less welcome, but even with these Nelson was gentle. Peter describes an impressive mousetrap Nelson made. "He built this elaborate castle out of wood. It was about four feet tall and had a winding staircase up, and it was a mousetrap. The idea was that at the very top of the staircase was a string with a piece of cheese on the end that the mouse couldn't reach. So the mouse would go winding around the stairway all the way to the top, jump for the cheese and fall into a five-gallon bucket. This thing was beautifully carved; it was lost in the fire, of course. And when the five-gallon bucket was full, he'd take the mice down and he'd let them go on the beach." Peter laughed, "I believe they would come right back in again."

---

[1] John T Walbran, *British Columbia Coast Names*, 1909

[2] Letter to Jim and Sarah Badke, May 2, 1988

[3] Letter to Pat Rafuse, December 12, 1982

[4] Letter to Leona Dolling, October 2, 1994

[5] Letter to Pat Rafuse, May 12, 1982

Nelson with Snuggles as a very young pup.

34. Creatures of the Sky

Speaking of birds, seagulls, shags, eagles, ducks—in the bushes down by the cabin there are many little birdlets no larger than walnuts[1] *and how they do twitter, twill and sing, 'This is the day, this is the day that the Lord hath made, we will twitter and twill, we will twitter and twill and be glad in it, and be glad in it.*[2]

Nelson delighted in watching the many birds on Copper Island, both large and small. In contrast to the walnut-birds, there was Ichabod (alternately remembered as Spindleshanks), the great blue heron, who fished from Nelson's float or along the shore, moving so slowly he was almost standing still on his spindly legs, then darting like lightning to grab a passing minnow and gobble it whole. Nelson II told me, "Ichabod used to come down and fish on the wharf, standing there, and my father would go out and sit with him and talk to him. And sometimes my father would walk down to the float and sit there, and Ichabod would see him and come flying in to stand beside him."

One time, Russell came up to me holding in his hand a tiny hummingbird that looked quite content there until it suddenly flew away. Patty told me a similar story about Nelson: "He had a bunch of hummingbirds that always came, but they sometimes got caught behind his veranda window. Nelson would gently take them out, then let them go. But one time, there was a hummingbird that didn't want to go. It sat on his finger, and he was just waiting for it and talking to it.

"And he thought, well, it's not leaving—I'll take him for a tour of the house. So he went upstairs with the hummingbird on his finger, and he showed it all the rooms and chatted away to the bird the whole time. He came downstairs and gave the hummingbird a tour of the entire house. It was just sitting and listening to him. He came back to the door and said, 'Well, it's time; you can go now if you want. We already had a tour of the house.' And off it flew."

Nelson II remembers Henry the eagle, one of Nelson's best friends. "Henry took hold of a fish in the bay, got swamped and needed to be rescued. So Nelson took the canoe out and somehow tied the eagle up so it couldn't attack him. He picked him up, brought him ashore and let him loose. And he called him Henry. The eagle befriended him and came every day to see him. Henry lived in a tree on Clifton Point and would constantly come and be with my father. When Nelson left the island, he bemoaned Henry and was troubled about how Henry might be doing."

Ben remembers a time on Copper Island when he heard what sounded like cats mewing in the trees. "I thought, that's weird. I didn't know there were cats here and I didn't know they would be up in the trees. So I looked up and it wasn't a cat at all; it was a crow." It's not so unusual for crows to imitate the sounds they hear, and they seem to use them intelligently. My wife and I used to encounter a certain crow when we took our dog for a walk in Lake Cowichan. Every time we passed by, it barked at us.

---

[1] Probably bushtits - audubon.org/field-guide/bird/bushtit

[2] Letter to Pat Rafuse, December 12, 1982

35. Creatures of the Sea

Nelson was once a Fisheries Officer, and his front yard was one of the finest fishing grounds in the world. Yet I never knew him to go fishing. He gladly received the salmon, halibut, lingcod, prawns and crab (and don't forget the abalone) that others gave him, especially if you also cooked or smoked it for him. I have wondered if, with his gentle heart, Nelson did not want to harm the depleted fish stocks any further, though he would not criticize his friends for doing so.

Nelson II told me the fascinating story of Matilda, the basking shark. Averaging seven to nine metres in length, basking sharks are the second-largest fish in the world. They are slightly smaller than whale sharks, and both species are filter feeders, their wide-open mouths straining plankton through enormous gills. Historically, basking sharks were plentiful on the West Coast, particularly in Barkley Sound.

"There was one big basking shark that used to come into the bay and rest up against the float," Nelson II remembers. "And my father would go down and pat her and talk to her, and he named her Matilda. Every day, Matilda would come and see him, and sometimes she would go out on the beach and rub her barnacles off. Nelson befriended her, and she would come and lean against the float and let him brush off her barnacles with his boat brush.

"And one day, Matilda never came back, and he bemoaned Matilda." In the 1940s and 50s, fishermen increasingly complained of basking sharks fouling their nets. "From 1955 to 1969, Basking Sharks in Barkley Sound were subject to an intensive and successful government eradication program. At the same time, a commercial fishery for shark livers, aggressive sport fishing, and general harassment of the sharks must have also contributed to their decline."[1] They are now an endangered species and are rarely seen in Barkley Sound. Fishermen at the time often attached a large, sharp blade to the bow of their boat, and that was likely Matilda's fate.

Basking shark, with divers for scale (Wikimedia Commons, public domain)

Nelson was concerned about how the fisheries were regulated in Barkley Sound. He told Patty that when he was a Fisheries Officer in the 1960s, many of his fellow officers were dishonest. "He said they shut down areas that shouldn't have been or let people into areas they shouldn't, so he quit." He had good reason to be concerned, as we now know. A news article in September 1976 celebrates a record catch for commercial sockeye that season. "Since the beginning of the fishery in May, over 680,000 sockeye have been caught, second only to the Fraser River catch, and three times the catch of the previous record year of 1973." In one week alone,

193 gillnetters and 98 seiners harvested 170,000 salmon.[2] Since then, salmon stocks have declined sharply, a problem intensified by climate change as the Pacific Ocean warms up.

Bill Irving remembers that the scraps from Nelson's kitchen went off the end of the dock and attracted wildlife, fish and seals. "When we first went down to Copper Island, there was fabulous lingcod fishing off Clifton Point on the Bamfield side," he told me. "But year after year, the cod diminished as more and more people tried their hand at it. So Nelson was somewhat concerned. We talked a few times about that. 'Some of those folks don't manage these things very well,' he would say. 'Do they really need that many fish?'" In July of 1984, Nelson wrote to Ron, "Well the 300 seiners and 600 gillnetters have pulled out and now the water is aboil with playboats. Speed rules the waves."[3] A couple of years later he wrote, "The ocean is full of play-boats agitating the water today. Poor fish—i feel sorry for them."[4]

In the eighteenth century, explorers noted that whales were abundant along the West Coast. Barkley Sound was especially known for orca and humpback whales, but not far offshore were Sei, Fin, Sperm and the massive Blue Whale, the largest creature on earth. The Huu-ay-aht people of Barkley Sound were known for whale hunting. Nelson was a friend to the Happynook family, whose grandfather Bill was one of the last to hunt a whale, back in the 1920s. A whale hunt was a serious endeavour for this community. It required physical, mental and spiritual preparation a year in advance and included ritual and ceremony. Whale hunters were honoured in the community, and when a whale was divided among the families, the most sought-after pieces went to the hunters.[5]

In the early 1900s, whaling stations were established all along the coast of Vancouver Island, including the Sechart Whaling Station near Copper Island. Between 1905 and 1917, this station processed 3000 whales. Two-thirds of those were humpback whales and over 100 were blue whales. I have been to the site of the whaling station, where Broken Islands Lodge[6] now welcomes kayakers exploring the nearby Broken Group Islands, part of Pacific Rim National Park. Behind the lodge is a small hill under which a huge pile of whale bones is buried, and harpoon heads have been found on the beach.

As a result of this slaughter, Nelson would have rarely sighted a whale from his window at Copper Island, and most of these would have been orca that sometimes chased a school of salmon into his bay. He may have also occasionally sighted a grey whale grabbing a snack after exiting off the migration highway along the West Coast from Baja Mexico to Alaska. He never mentioned whales to me or noted them in his letters.

However, a drama was unfolding on the other side of Copper Island. Sealand of the Pacific was an infamous public aquarium at Oak Bay in Victoria. Due to the deaths of two of its orcas, they obtained permission in 1983 to capture two orcas from L-pod, which ranged along the West Coast. A huge protest ensued, involving conservation groups such as Greenpeace and Islands Trust. The activists converged on Holford Bay on the west side of Copper Island, where Sealand had set up capture nets. Sealand eventually gave up, taking three orcas from Iceland instead—one of which killed an aquarium worker, causing the site to be permanently closed.

In recent years, humpback whales have returned in significant numbers to Barkley Sound. Nearly every time I have visited Copper Island in the past ten years, I have sighted at least one humpback as it passed by the island. I was once standing on the float when, not 20 metres away, a humpback rose out of the middle of the bay with its mouth wide open, half-breaching out of the water and collapsing back with a huge splash. Sea lions and seals are also increasing in number, provoking a potential showdown with those who are trying to preserve the salmon stocks.

One last creature of the sea deserves mention, not because Nelson ever talked about it but because of the name it was given. In the early 1960s, fishermen and the crew and passengers of *Lady Rose* began reporting sightings of a strange monster in the waters of Barkley Sound. Similar reports had occurred in the 1940s when the creature was named "Tzartusaurus" as it was sighted near Copper (Tzartus) Island. It was said to be between 20 and 40 feet in length and of a colour and appearance unlike any whale or fish. Some identified it as an elephant seal, but others who claimed to have seen the creature downplayed that suggestion, saying it was more like descriptions of the Okanagan's Ogopogo or Victoria's Cadborosaurus.[7]

Alberni Inlet Monster

'Tzartusaurus' Seen

PORT ALBERNI—"Tzartusaurus" has made an appearance in Alberni Inlet.

A monster with qualities comparable to those of Vic- Alberni Marine Transportation Ltd. vessel Lady Rose.

Although he has never seen the creature in its entirety, John Monrufet of Marine Transportation said he is con- hontas Rock and Chup Point. The latter point is close to Tzartus Island. Kildonan residents have also reported seeing Tzartusaurus and one resident of the village said he ob- make an appearance for a period of several days.

The turmoil made by the creature in the water has impressed most of the travellers.

While Mr. Monrufet has [8]

---

[1] "Basking Shark," *Fisheries and Ocean Canada*, dfo-mpo.gc.ca/species-especes/profiles-profils/baskingshark-requinpelerin-atl-eng.html

[2] "Commercial sockeye record catch this year," *The Daily Colonist*, Sep 8, 1976

[3] Letter to Ron Pollock, July 22, 1984

[4] Letter to Ron Pollock, August 23, 1986

[5] "The lasting legacy of Nuu-chah-nulth whaling off western Vancouver Island," *Chek News*, September 7, 2023, cheknews.ca/the-lasting-legacy-of-nuu-chah-nulth-whaling-off-western-vancouver-island-1167794/

[6] Broken Islands Lodge, brokenislandslodge.com

[7] "Tzartusaurus: West Coast 'Serpent' On Prowl," *The Vancouver Sun*, Feb 2, 1961

[8] "'Tzartusaurus' Seen," *The Daily Colonist*, February 2, 1961

These are two stanzas—illustrated by Nelson—of the poem, "Tis the Set of the Sail," written by Ella Wheeler Wilcox in 1916

36. The Mine

When Captain Edward Stamp formed the Barkley Sound Copper Company in 1863, he didn't know that the real wealth on Copper Island was actually iron in the form of high-quality magnetite. Iron claims were first staked there in 1894, and they were worked at the top of the mountain above Clifton Point until 1902.[1] Several open cuts were made that can still be seen today in the undergrowth. More magnetite was extracted from the ground than was transported down the mountain and off the island. "For over twenty years some two thousand tons of magnetite of commercial grade has been lying on the dump on one of these claims, 900 hundred feet above the sea," said a 1922 news article.[2]

At the same time, Clifton Point was also explored. Especially at low tide, the rocks and outcrops of the point are stained red with oxidation. "On a small neck of land on the east side of this island, near a good sheltered anchorage, a shaft has been sunk 50 or 60 feet on a ledge of magnetite that carries more or less sulphides. This ledge out-crops irregularly along the shore, some parts very red or reddish brown proving on fracture to be solid pyrites. There is a good house near the shaft, but all the men had been removed to the Sarita River property."[3] Aaron told me that he was astonished when he first saw this shaft on Clifton Point—a great gaping darkness that dropped down forever—or at least a great distance below sea level. The shaft has been carefully covered and made inaccessible.

Though surveyed several times, no other mining took place on Copper Island until 1961, not long after the Dunkin family arrived. The Empire Development Co. Ltd. "optioned one recorded and seven Crown-granted claims on a steep hillside above Clifton Bay on the east coast of Tzartus Island... Dip-needle surveying and 2,270 feet of diamond drilling indicated a triangular-shaped layer of magnetite lying at or near the surface and not more than 25 feet thick... An average of three persons was employed from January to mid-July [1961]."[4] Since there is no further record of ore removal, the deposits must have been deemed not worth the effort.

Nelson II remembers the miners building a diamond drilling camp on the Dunkin property. "They set up temporary camps right there, and the diamond drillers were working there. During that time, my mother was the cook. I used to help her cook for the crews, and I worked with them. They were a crew out of Quebec, and they had tents with platforms as their bunkhouse, up on the hill between Nelson's house and Pebble Beach. They also set up a cookhouse. The engineers stayed in the cabin down the west shore of the bay."

To my knowledge, Nelson never considered working the mine himself, though he was interested. In a letter to Ron, he wrote, "So the mineral claims here are still being held. Thanks for finding out."[5] Visitors to Copper Island sometimes make the steep trek up the mountain to the single mine shaft near the top. Strewn at the mouth of the mine are large chunks of magnetite, surprisingly heavy and able to stick strongly to a magnet. Water usually pools in the mine entrance, and I have never ventured far inside. I don't mind caves, which have been there for centuries, but any mine shaft in BC is a relatively young thing and not to be trusted.

Ben remembers, "I decided one day to go with some friends to the top of Copper Mountain. As we found our way up, my dog chased a deer down the trail we were coming up.

The deer was terrified with a dog behind it, so it didn't worry about us. It passed within two or three feet of us. As it roared past, if I had set my foot out, I probably could have drifted it. We reached the top where the mine was located and walked into the shaft. At the end of the shaft, I found a felt cowboy hat with what looked like a bullet hole in the front. That piqued my imagination and I wondered, what was the story behind the hat?"

The flooded magnetite mine shaft near the top of Copper Mountain.

---

[1] Kirk D. Hancock, *Magnetite Occurrences in British Columbia*, 1988

[2] "The West Coast of Vancouver Island," *The Daily Colonist*, October 8, 1922

[3] William A. Carlyle, *Report on the Alberni Mining District*, 1896

[4] Ministry of Mines and Petroleum Resources, Province of BC, *Annual Report,* 1961

[5] Letter to Ron Pollock, February 21, 1983

Candlestick turned by the author on Nelson's foot-powered wood lathe.

37. Faithful Tools

Nelson loved and depended on his tools. Eventually, most of his tools were manually operated. He had some gas and electronic tools, but he always ended up giving them away, especially after he donated his generator to Coastal Missions. Brian told me, "When I bought my big fishboat, the *Kolberg*, it came with a hand plane that the fellow had used to smooth the planks cut with his power saw. The plane was really long and I had no use for it, but Nelson wanted it. So he traded me his electric plane for it since he had no electricity. I bought his power saw too, one with a very long bar."

As generous as Nelson was, he was a salvager of everything he could find or that came to him. Rich remembers that Nelson noticed boaters leaving a bit of oil at the bottom of each quart container. "Goodness, they were sixty-nine cents a quart. What were they thinking? So he made a rack and set these 'empty' containers upside down on it, and all the oil dripped down and funnelled into a container at the bottom." Nelson salvaged several quarts of oil that way, at a great savings!

He appreciated and found a use for all the odd bits people brought him. He wrote to Heather, "Dear old Ron is so kind and helpful, always giving; he has helped me much with the things he gives me. So what if it be things others throw away? Bent nails are as good as new when on my anvil i get through with them. Screws and bolts and pieces of machines are all useful to me."[1] To Ron he wrote, "i put everything to good use and if it is not needed now i stash it away, and then someday it will be the very thing i need."[2]

Nelson at his homemade metal lathe and grinding wheel in 1987.

37. Faithful Tools

Of the tools Nelson made, people remember his lathes the best. When he lost his original lathe in the fire, he made two more, both operated by foot treadle. One was a metal lathe that also had a grinding wheel, and the other was a lathe for turning wood. He created both with parts he had salvaged or ordered by mail, and both were precision-made and very solid. He wrote to Pat, "Today it was raining so i stayed in the shop. Finished making up a belt to hold me from slipping back on the seat while peddling the lathe, and also turned a piece of alder."

I loved working on his wood lathe. It sat beside the window with a good view of the bay, and no motor noise spoiled the effect. There was something satisfying about the effort needed to keep the spindle turning while concentrating on the cutting tool's shaping of the wood. I made a candleholder, which required a round base with an attached candlestick that I carved after turning the wood. Other visitors found the same fascination in shaping unique wooden pieces on that lathe.

If Nelson needed a tool, he often made it himself, though he also salvaged some and bought others from mail-order catalogues. Nelson II said, "If he was doing a job and he needed something, he'd make it. He made all kinds of homemade tools and appliances. He even made his own padlocks and other sorts of locks. When I was a kid, he built a writing desk and made the brass locks for it. He also made very advanced brass locks for our home and other places."

Other innovations included Nelson's large post drill. He wrote, "i am still working away at the big 2 speed post drill and hope to have it running before long now."[3] The drill stood six feet tall and sat in a corner of the shop at Copper Island Camp for many years, but was lost in yet another fire that happened there several years ago. This drill was hand-cranked but had a large flywheel to maintain momentum and provide good torque for drilling.

Bill Priest was especially fascinated by Nelson's blacksmith shop. "I thought that was pretty cool because there sure weren't many of those around. Nelson was a guy who knew how to work metal. I enjoyed watching him." Nelson II agrees: "My father was a tremendous blacksmith—he could make anything in the blacksmith shop."

"Everything in its place." Nelson's blacksmith shop.[4]

The tools are all gone now—lost to the fire or distributed among his many friends. Some of his tools were donated to the Alberni Heritage Museum[5] as rare pieces of craftsmanship. Even more rare these days is the kind of tenacious ingenuity that Nelson demonstrated in creating much out of nothing.

---

[1] Letter to Heather Arnott, November 23, 1986

[2] Letter to Ron Pollock, undated

[3] Letter to Jim and Sarah Badke, November 1988

[4] Bamfield Community Museum and Archives

[5] You can take a virtual tour of the museum online—search "Alberni Heritage Museum"

38. Messing About in Boats

When you live on an island, boats become a necessary part of everyday life. "In Bamfield, you're not a one- or two-car family," Ben chuckled. "You're a one- or two-boat family." Of all the boats that Nelson depended on, none was more critical than MV *Lady Rose*, the freight ferry that also carried passengers and ran between Port Alberni and Bamfield. A few years ago, the company decided it was time for them to close, and a great cry went up from the people of Barkley Sound, as well as from the many visitors who loved a leisurely Saturday cruise on the Alberni Inlet. Fortunately, locals Greg Willmon and Barrie Rogers of Devon Transport bought Lady Rose Marine Services in 2021, keeping the service alive with the current vessel, MV *Frances Barkley*.

The MV *Lady Rose* approaches Nelson's float, having come around Clifton Point.

MV *Lady Rose* (christened *Lady Sylvia*) was, in 1937, the first single-propulsion vessel to cross the Atlantic. On its journey to Canada, "it was soon apparent that the *Lady Rose* was a lively ship with a movement which made it difficult and often impossible to do more than hold on grimly."[1] Upon arrival in Vancouver, the boat serviced several ports—and the war effort—for many years before arriving in Port Alberni in 1960, shortly after the Dunkins moved to Copper Island.

Lady Rose carried "mail, passengers and cargo back and forth from the pulp and paper mill town of Port Alberni to the small fishing villages of Bamfield and Ucluelet in Barkley Sound and any required stops in between. For those living on the many islands which dot the Sound, the ferry is their lifeline to the outside world. [Captain George] Monrufet says the crew of the *Lady Rose* once did the grocery shopping in Port Alberni for one of the more hermit-like island residents."[2] This was probably Nelson.

Patty remembers that when they saw *Lady Rose* coming up the channel in the distance, "if you saw the 'rabbit ears' turn straight, you knew it wasn't coming in. But if it angled the other way, you could tell it was coming in." Nelson wrote to Ron, "Al will try to flag the *Lady Rose* on their way to Banfield as they have a much greater chance to see him than when going back." The passenger ship was Mina's connection to family and community off the island, and she made the trip whenever possible.

When the Dunkins lived in Kildonan, they had several boats, as famously told to you by R. Bruce Scott in a previous chapter. Nelson had a fascination with cedar dugout canoes, having first bought one from Lloyd Bridal's family before leaving Port Albion. Nelson II remembers, "He felled a big cedar tree and built a dugout canoe, which we used for years and years. We used to have an outboard motor on it, and he did other boats, too. He was a boat builder. That's what he did." Peter remembers this dugout as being part of the story of the Dunkins' Christmas adventure.

When they were living on Copper Island, Nelson's most memorable boat was the MV *Raven*. Nelson II recalls, "The *Raven* was built by a man named Bill Fowler in the Somass River area of Alberni in 1933. I remember that because the boat was 33 feet long. Somehow, my father bought *Raven*, and we used that for many years." It was "an object of much tinkering, with its 2-cylinder Easthope engine."[3] Bill Irving told me, "I remember several times when he was getting older, him saying, 'Come on, Bill. Come down here and help me start this thing.' You had to kneel in front of it and grab the big flywheel and crank it over until there was some compression, and then give it a good boost."

Raven tied up alongside Nelson's float and hoist.[4]

Debbie remembers the boat fondly. "One time, Nelson took me out in *Raven*. And it had this Easthope engine that went kathump, kathump, kathump. This was in Bamfield, and we were on the Marine Station side, and he was going to take me to the grocery store on the other side of the inlet. And he had these little wire glasses in the case, and he put them on, and he said to me, 'Do you want to go fast or do you want to go slow?' And I said, 'I want to go fast!' There was no change in our speed, but he was so happy." Roy said, "You could hear Nelson coming to town miles away with the *Raven*. It was unbelievably slow. Two-and-a-half knots or so."

When Brian was at his mother's place doing projects, occasionally Nelson would come by with *Raven*, towing a log he had found adrift on his way through Trevor Channel. "He would hook onto it, and instead of taking it home, he would bring it, knowing we would need a log for firewood. It's 6 miles. He would come in, swing by and stop at the float for a few minutes. Of course, it took him at least an hour or so to get there."

Eventually, Nelson II and his dad rebuilt *Raven*, and the son began using it as a fishboat, trolling out of Bamfield. As an 18-year-old, Nelson II lived on the boat for a while. But it was an ancient boat with many problems. "The *Raven* kind of gave up," Roy told me, "and I did know Nelson enough to say, Listen, I'll build you another boat. But he never took me up on it, even though I could have. The boat that he needed was pretty simple to come up with."

Raven in Nelson's bay, doing what it was most inclined to do: sink.

Rich said, "The first time I went to Copper Island, the *Raven* was sitting there. She was black, and I admired it. The boat was very shapely, a dory-type, and I said, 'Oh, I like your boat.' He said, 'You like it?' I said something like yes. He said, 'Well, no, I would give it to you, but I would only give that boat to my enemy. I would never give that boat to my friend.' He said it was a boat that was determined to sink. 'Yeah, I like you too much to give you that boat.'"

Bill Irving remembers that *Raven* finally did sink, right in Nelson's bay. "He didn't have the energy or capacity to bring it back to life again. And you could see a bit of the energy leaving him, knowing the restrictions of having no transportation out other than *Lady Rose* or somebody coming down."

I asked Brian what happened to the boat. "The old *Raven*?" he replied. "Yeah, I burned it up. It sat on the beach for a long time. And one time he said, 'Could you help me? We need to burn this up and get rid of it.' So that's what I did. He was not young anymore, and there wasn't much he could do with that old boat."

Ink sketch of *Raven* at Nelson's float, by Roy Getman

---

1 "Brave little Lady crossed Atlantic," *Times Colonist*, March 25, 1990

2 "Lifeline to the Outside World," *The Daily Colonist*, September 8, 1976

3 Adele Wickett, "Remembering Nelson Dunkin," *Island Christian Info*, June 1998.

4 Bamfield Community Museum and Archives

39. The Cedar Dugout Canoe

Not that Nelson was done with boats. Everyone remembers the years that he spent building a dugout canoe to sail around the world. The idea was not as far-fetched as it may seem, though such a feat was far beyond Nelson's capacity at his age. In 1901, Captain John Voss bought a dugout cedar canoe from a Nuu-Chah-Nulth village and set out from Victoria. Over a period of three years, he sailed the "*Tilikum*" to Australia, New Zealand, South Africa and Brazil, and finished his journey in England.[1] This was Nelson's inspiration to build a canoe solid enough for such an adventure.

Nelson did not think well of John Voss. "Right well i know Clifton Point but Voss Point i am not acquainted with. i presume it is at Dodger's Cove from which Voss departed on his infamous abortive voyage around world," he wrote. "Voss was a wicked man. In his book, he made a joke that he had broken the law by bribing with whiskey the Indian from whom he got the *Tillicum*. From Victoria, he went to Dodger's Cove from whence he departed after some days of boozing with the store-keeper and robbing the Native graves for skulls with which he wanted to make himself some great attraction to the foolish."[2]

Ben remembers, "Nelson, a man of practical faith, would always thank God if a storm blew a log into the inside waters of Clifton Point. A storm brought a large cedar log into Nelson's front yard. After hauling it up well above the high tide mark, using come-alongs, Nelson began the huge job of trying to turn this log into a sailing vessel." It was very narrow, and he built a cabin on it and sealed the top so the high seas wouldn't sink it. Patty said, "He always told us he was going around the world in the canoe, and we would always chuckle to ourselves. I think it was one big joke because we kept teasing him about it, and he'd tell us exactly how he was going to do it. He just played along with us."

A full 46' in length, Nelson's dugout canoe was an astounding project at his age.[3]

Peter came over from Kildonan to mill some wood for the canoe. "Nelson was going to saw the canoe in half and make it wider. He had a creosote piling that we squared up for him with his Alaska mill, and he was going to bolt it in between to give the canoe a wider beam." Peter also remembers the brass bow plate Nelson fashioned from "an inch and a quarter bronze eight-foot propeller shaft out of a fishboat. He chucked it up in his vise, and every day as he walked by on his rounds, he grabbed the hacksaw and took a few swipes." Peter thought it might have taken Nelson a year to cut the parts he needed.

When I saw the dugout, I prayed Nelson would never launch it into the water. I said to Brian, "I'm really glad he never finished the canoe. I don't think it was seaworthy." He replied, "I don't think so either. Yeah, Uncle Roy had Nelson figured out years before. Here's Nelson working on this, and time goes on, and he's still working on it. And working on it. Roy says, 'Nelson doesn't really want a boat; he just wants to build a boat. He's never going to finish it." Bill Priest agrees: "Nelson didn't have a boat and never wanted to get off the island. It would have been fun to see if the dugout would actually get out on the water and do what it was supposed to do. But it was not to be. I don't know how seaworthy it would have been. It was quite narrow. But who knows?"

It wasn't Nelson's only attempt to build such a craft. Rich tells the story of another edition he encountered: "When I was over there one time, I saw this bizarre boat made out of a log, a 30-foot dugout. Nelson had read with interest the story of the *Tilikum*, the dugout that sailed around the world. So he had built this boat from a hollowed-out log. It had a keel and a full cabin on it. You crawled in through the back.

"I took an interest in it since I had also read the story of the *Tilikum*, and I said, 'Wow, this is an amazing boat!' He talked to me for a while about the boat, and he said, 'Well, do you want it?' I don't know what I stammered. I don't think I said, 'Yes, I'll have it.' But he thought I did. One day, I looked out the window of my house, which is on the waterfront in Ucluelet, and I saw a boat coming down the harbour, the

Nelson E. Dunkin, the one that was rebuilt. And it was towing this very strange craft.

"Don told me, 'Nelson said that Rich Parley wanted this boat. Could you take it over to him?' There was Don out on the water in the *Nelson E. Dunkin*, and I don't know where we found an anchor, but we had something, and there the thing swung at anchor. What am I going to do with it? Storms came and went, and everybody had to stop by to see this dugout canoe. So I made a wooden sign for the window on the canoe, and it said, 'Unique craft. Twenty-five cents a look. Just throw the coin inside.' One of those quarters was in the bilge for a long time.

"Finally, I didn't know what to do with it. Some native people came by, and they said, 'What are you going to do with that boat?' I just quoted Nelson Dunkin, 'Do you want it?' They took it away, and it came to a sad but picturesque end. They were using it for gooey duck harvesting, as they felt it was tough enough to bump into the rocks on the swells and survive. It didn't. The boat came up on a big swell, came down on top of the reef and broke in half, and somebody had to rescue them."

With the dry weather i've been working on the canoe and have spliced it out to 46 feet and turned it right side up.[4]

i have managed (in spite of the swarms of mosquitos) to get considerable done on the canoe.

Monday Evening. Another fine day. Finished putting up the four cabin corner posts on the canoe and am as tired as though doing a day's work. Lord willing and with his help someday we should have a little boat.

Tuesday evening after a fine summer's day. This morning showed up Ben Potter, his son, and Rev. Buchanan of Kamloops. They helped with the canoe and so got another side of ribs in, towards the bow: looking more like a boat all the time.[5]

Well yes, i do keep busy with numerous projects. Last week i was back at working on my big canoe, chopping out a stern piece.[6]

The sternwheeler Tipperary is beached for the time being. As a boat, it is a failure but as a lesson, it is first-rate. i found that the forward motion was not sufficient to keep the wheel going over the 2 dead centres (that is with lever action) without undue attention on the lever. Also i learned that if the paddlewheel is too deep in the water the action is poor. To work right, i would have had to incorporate a flywheel and that would have taken until Christmas so i pulled the paddlewheel off, put it in storage and set the ship ashore. Also the hull is such an ill-shaped thing that there is no hope for it except to double its length.[7]

i just keep on keeping on and wondering whatever the Lord can do with me. The little 12-foot scow is progressing and (Lord-willing) should be ready for launching by next month. But woe-is-me, i have neither the jug of Old Irish Bogwater or the Fair Colleen to smash it over the end that goes first.[8]

Nelson's other "dugout" was the handheld toy that he wished he had never made, as people wasted so much time with the "silly game." It is described as containing either a pea pod or a miniature dugout canoe, which players had to shake until they got all the peas (or people) into the pod (or canoe). Now owned by Pat Rafuse.

---

[1] John M. MacFarlane and Lynn J. Salmon, *Around the World in a Dugout Canoe: The Untold Story of Captain John Voss and the Tilikum*, 2019

[2] Collected poems and prose of Nelson Dunkin, given to Jim Badke

[3] Bamfield Community Museum and Archives

[4] Letter to Jim and Sarah Badke, October 25, 1986

[5] Letter to Jim and Sarah Badke, August, 1987

[6] Letter to Ron Pollock, undated

[7] Letter to Don and Patty Cameron, January 3, possibly 1982

[8] Letter to Ron Pollock, July 22, 1984

40. The Toy Works

The Toy Works was a dream of Nelson's that never materialized. It was his answer to how a community of people residing on Copper Island could support themselves while creating simple, wholesome toys for children, unlike the ones he lamented when he scrutinized the Sears Christmas Wish Book. I will let him tell you about his dream in his own words:

Have you noticed the toys in the catalogues? Outright demonic. So if a young child is brought up on such things what must we expect them to be when grown? Now Copper Island Enterprises Wooden Toy Division aims that all toys will be of a kind and loving nature—such as for the very young, a Waddly Duck which when pulled along by a string (no, not by a costly battery) she will waddle along from side to side and it could be if plans were made with simplicity and long life in mind, she could be made to say, "Quack Quack!" And for the little girls from age 3 to 30, there is the ever popular Doll House.

Now about your father making wooden toys: would he make them to sell or give away? When the TOY WORKS is built at the waterfall would he be interested to work there? Either way he could make his toys to sell or to give. The Toy Works will be small but could be run 24 hours a day and there is an abundance of good Alder Toy Wood. Also, there are many other wooden things that could be manufactured. For the things made for money, we would need to look into the best mode of marketing so as not to be working for the big stores at a pittance. It would be well to check with the Salvation Army stores.

i have been enquiring around about buying a small lathe but as yet no success. From the lathe (12" x 3') that went through the fire, i have the headstock in the shop and have started taking it apart—it might just be that it can be made to work—that is simple, not using the gears except the back-gear. It is quite necessary in building a Toy Works to have a lathe that will turn shafts and bore brushings. Also, i am in the market for any old machine that no one wants anymore—now, all the machines are coming out computerized. To date i have the little lathe which i built, a 9" Craftsman grinder-sander and a Makita reciprocating saw. Before any great thing can proceed, i must build a British (Scotch) saw bench to saw lumber on.[1]

Concerning Shop: Get the waterwheel and shop built over at the waterfall and we could work day and night. Of course, most of my work would be the toys for poor children. Something i think would sell: a cedar sailboat hull turned out in the rough by a carving machine; this would be a hobby kit, the hull and a bundle of split cedar to make the masts and to finish off deck and cabins, perhaps 20 inches long. The thing is to get a way of marketing them without some middle-man parasite getting most of the money. Perhaps Christian agents would be the way.[2]

Over at the creek and Toy Factory i hope to keep the building and cabins rustic and plain—burn wood and suffer by without TV. With the power and a few basic machines, i am sure you could turn out some very salable products.

If in your travels you come across someone who knows about paints (that is the making of paints) perhaps they might consent to tell you if there is a wood stain can be kept in a vat so parts of toys can be thrown in—swished about—fished out and dried (nontoxic i suppose) as there seems much concern that children will eat toys and be poisoned. Also, should any nails and screws be made of rock candy?

Monday morning: Cold and wet: Here is an idea for you to think upon: If it seemeth good to thee—get at making: if it seemeth to thee of no worth then forget it as an evil thought: 1-pair children's skis : 1-minature cutter body : This cutter body is size to fit the child : a comfortable seat and all lined with fuzzy material with a hood to pull up so inside it is weather-proof, also it is equipped with straps so it also becomes a backpack : The skis are made so as to be clipped on under the box and thus you have an all-around weather cutter to pull the kiddo in : Take the skis off the box and your little one has his own skis : In the summer you could make this (possibly in three sizes?) and in the winter sell them.[3]

I did take Nelson's suggestion and built a sled for our kids that I could pull behind me on cross-country skis. But I couldn't find a way to make the skis detachable from the sled. Nor did I—or anyone—ever go to work at the Copper Island Enterprises Wooden Toy Division.

[1] Letter to Jim & Sarah Badke, October 25, 1986

[2] Letter to Jim & Sarah Badke, April 1987

[3] Letter to Jim & Sarah Badke, February 1988

Mary Scholey: missionary, postmistress and a faithful friend to Nelson.
Photo: Bamfield Community Museum and Archives

41. Mary Scholey

i think sometimes that if it wasn't for Mary, i would have died some time ago."[1]

I met Mary just once. My wife and son and I were visiting Nelson for a few days, and on the Saturday, Nelson looked out the window and said, "Well, now, here's Mary!" We followed him to the float, where a smaller, older woman in a light blue Mustang flotation jacket was deftly mooring her 15-foot outboard. We walked back up to the house, and she unloaded the things she had brought—a few groceries (including bananas, of course), Nelson's mail, treats for Snuggles and a container of ice cream.

The first order of business was to polish off the ice cream, which was already more cream than ice. Mary and Nelson talked about happenings in town, the news, projects Nelson was working on, and mostly about us, his visitors. Mary didn't stay long, which Nelson told us was always the case. She would usually also cross over to just inside the Sarita River to do the same with Angie Joe and her son Wilson, of whom Nelson once wrote, "They too would be in a sorry plight if it wasn't for Mary looking after them." Plus, Mary always kept an eye on the weather and didn't want to be caught out in the channel past dusk.

I once heard a story—and I cannot verify the source—but even if it is not true, the tale exemplifies Mary's practical faith and resilience. Mary went out in her boat one afternoon and—whether she ran out of gas or something was wrong with the motor—found herself stranded in the middle of the channel.

Tide and wind drew her to the mouth of Barkley Sound and out into the open sea, out of sight of land. She drifted, so the story goes, for two days, living on the groceries she had planned to deliver and communing with the God who made land and sea. Finally, a passing fishing boat spotted her and brought her back to Bamfield.

2

In between serving customers, sorting mail and handing out biscuits to the neighbourhood dogs, Mary Scholey leans on the dutch door of the Bamfield post office, enjoying the view from the waterfront boardwalk across the inlet that separates the two halves of the village.

As postmistress at this tiny outpost—with a winter population of about 350 people—Scholey works four days a week, sorting sacks of letters and magazines dropped off by the mailboat. Her most unusual packages are the boxes of live worms destined for the Bamfield marine station. During the summer, she is kept busy with tourists who consider the Bamfield postmark a hot collectible.

Mary has been running the Bamfield post office for about 15 years. She is well past retirement age, but finds the post office pace ideal. "It's never quiet as far as I'm concerned," said Scholey, who still runs Bible classes on Sundays.

She also refuses to divulge her age. "I never tell people how old I am," she said, her face creased with laugh lines. "When I was a girl, people told me I was too young to be in Bible college. Now they say, 'Oh, you're too old to be running around in a speedboat.'"

Would she ever retire to someplace like Florida? She shakes her head. "I wouldn't like to be in one of those same-all-the-time climates," she says, enjoying Bamfield's unusual winter sunshine.[3]

Mary came from her home in Saskatchewan to attend Bible college in Victoria. After graduating in 1948, she became a missionary with the Pentecostal Assemblies of Canada. In the 1950s, she and Marion Johnson established a Pentecostal mission in the Grappler Creek Reserve, where a church was built soon after. The community later moved to the Anacla village at Pachena River, and the church at Grappler Creek was abandoned. Mary continued to lead children's programs and Bible studies in the Bamfield area and also served at Copper Island Camp in its early days.[4] In 1983, she became the postmistress of the Bamfield Post Office, a role she continued for 23 years.

Mary's life was difficult and lonely, but it was the life she felt called to, and she was both faithful and courageous. Ben remembers, "She was a good boat handler; almost any weather, she'd come the six miles from Bamfield. And Trevor Channel can get fairly rough." Pat agrees: "She could handle the weather. She'd drive her boat, and she was a really resilient person. Mary was so faithful."

Mary had her setbacks: "Miss Mary Scholey is a patient at the West Coast General Hospital after injuring her leg while trying to start her outboard motor. Her leg was badly lacerated and bruised and after emergency treatment at the Outpost Hospital, she was flown to Port Alberni."[5] "Mary Scholey of Bamfield was charged July 27 with driving left of the centre roadway following a two-car accident about 10:55 p.m. on Franklin River Rd. There were no injuries. Damages totalled $5,500."[6] That would have been a trip over the notorious logging road to Lake Cowichan to serve at the camp at the West Coast Indian Fellowship on Riverbottom Road in Duncan. Faithfulness doesn't always mean smooth sailing.

2

Nelson looked forward to Mary's visits and became increasingly dependent on them. Here are a few excerpts from his letters that show how much she came to mean to him:

The winds were boisterous and the waves were rough, day after day, but finally Friday was a calm nice day and Mary came with the mail and 2 loaves of bread, 2 quarts of milk and a tray of cupcakes and also my money—she is very kind to do these things for me.[7]

Mary, with all the things she has to do, is always most helpful and prompt…[8]

Now the first day of November and wondering as usual if Mary will come… Came and ready to go.[9]

Resurrection Sunday. Mary did not come yesterday. This forenoon bitter cold—this afternoon calm and overcast. No one came except one yacht which ran into the cove so as to turn around and go back out. Feeling right useless today—have been working on my picture albums—now on the 4th one and still about as muddled as when i started. i guess about everyone will be going to church today, and the next time will be Christmas, will it not?

Monday. Mary here now, Praise God.[10]

Saturday: Much overcast but calm so will see if Mary comes. Mary arrived and in a hurry as usual.[11]

Mary never came yesterday and i presume because there was quite a brisk wind blowing. So i suppose it will be another week ere i see her, Lord willing and the weather calm.[12]

A Hundred Thousand Thanks for those Yum Yummy cupcakes. When Mary brought them i didn't know, as she had them down in the bottom of the bag of groceries she brings. Next time up we had them. Snuggles had a whole one to herself. There was one extra but Mary refused to take it—nevertheless, she did take it as i slipped it in her bag when she was not looking. She liked them.[13]

Since December 7th, Mary has had no boat as her motor went broke, a need for prayer.[14]

The sea was rough so Mary sent parcels from you by way of a fisherman. Whatever would i do without Mary? Lord bless her.[15]

Mary (Lord Bless her) is most kind towards me and always what more she can do for me, but though i love her i have a respectful fear of her.[16]

Rain without wind today and yesterday was calm without rain so Mary came after some days of southeaster winds and heavy rain. She went across and towed that old boat back from the fishing lodge, which we can fix up well enough here for beach combing.[17]

Mary sprained an ankle and was on crutches but thankfully is getting better.[18]

As she is wont to do, Mary came yesterday with mail and eaties. If it wasn't for Mary, i would have a hard time staying here. Mary badly sprained her ankle some time back doing good to people but now, thanks to Jesus, she is able to get around again, spry as a squirrel and happy as a wren.[19]

Patty remembers, "Mary took Nelson under her wing after Mina had passed away. She would come and visit regularly and bring the mail. And I know Nelson had a real liking for her because we had heard at one point that he actually would have liked to marry her. I think he even asked her. She didn't think that was a good idea, but she would still be his friend and bring him stuff. But yeah, what a lovely lady. We got to know her really well, and I enjoyed her visits when she came over. And then she had to go and get back to Bamfield. She always had something to do there, and she kept very busy. Nelson always looked forward to her coming."

Perhaps Nelson's conflicted feelings about Mary are summed up well in this letter to Pat: "Well, wise or foolish, i am doing another birthday gift for Mary. If i didn't do i would feel miserable and if i do, do i feel perhaps it would be as well left undone? Ah, miserable wretch that i am, why do i not go to the North Pole and climb it?"

20

SCHOLEY, Mary

Our dear, sweet Mary died Sunday Feb. 26, 2006 in Victoria, BC. Born in Saskatoon, Saskatchewan, June 23, 1920, the second daughter of Rose and William Scholey living near Netherhill Sask. Predeceased by her oldest brother Bernard ("Slim"), Mary is survived by her sister Rose Bursey of Victoria and brother Bill (Charlotte) of Vernon BC, their children and her precious dog "Dutchess". Mary graduated from the BC Bible Institute in Victoria in 1948 and then became an licensed minister with the Pentacostal Assemblies of Canada. She moved to Bamfield BC and served there as minister, Sunday School teacher, counsellor and friend to all. She travelled many miles in her speed boat taking groceries, mail and God's word to those who couldn't come to her. Mary was a well known fixture of Bamfield, BC and for the past 23 years was also Bamfield's postmaster, a job that she loved. Mary's last wish was, "Love one another just as God has loved you" and meet me in heaven.

A celebration of Mary's life will be held on Monday, March 13 at 2pm in Bamfield BC. Information: sjohnsonrn@shaw.ca

213722

21

But Nelson did delight in any opportunity to please Mary. He wrote to Heather—not from the island but from a faraway dilapidated apartment in Greater Vancouver—"But Oh, how i long to be working for the Lord making Scripture plaques. i have joyous news! Mary has asked me to make plaques for her windows. And i have written to tell her there is no one i would rather make plaques for. She says they attract many to the way of salvation."[22]

---

[1] Spotted in a letter from Nelson that Madge asked me not to photograph

[2] Bamfield Community Museum and Archives

[3] "Something about Mary," *Times Colonist*, March 25, 2001

[4] Pat Garcia, *Bamfield: Looking Back*, 2010

[5] "Around the Harbour," *Alberni Valley Times*, March 22, 1971

[6] "Police News," *Alberni Valley Times*, July 31, 1979

[7] Letter to Pat Rafuse, February 20, 1983

[8] Letter to Pat Rafuse, December 12, 1982

[9] Letter to Jim and Sarah Badke, October 26, 1986

[10] Letter to Jim and Sarah Badke, April 19, 1987

[11] Letter to Pat Rafuse, May 12, 1982

[12] Letter to Pat Rafuse, May 12, 1982

[13] Letter to Pat Rafuse, December 12, 1982

[14] Letter to Ron Pollock, undated

[15] Letter to Ron Pollock, February 21, 1983

[16] Letter to Pat Rafuse, December 12, 1982

[17] Letter to Ron Pollock, October 31, probably 1980

[18] Letter to Ron Pollock, September 1, 1985

[19] Letter to Ron Pollock, October 6, 1985

[20] Bamfield Community Museum and Archives

[21] "Obituaries," *Times Colonist*, March 11, 2006

[22] Letter to Heather Arnott, undated

42. Letters By Mail

Yesterday was a cool, dark calm day and Mary came with the mail and other things: Now to day a South Easter is blowing, the sea is a humping and the rain comes sloshing down. [1]

Imagine living in a home with no electricity. We can endure a power outage for a few hours, or go camping for a few days without lights or heat or technology—in the warm summer months when it's light out until ten o'clock. But what about in the dead of winter, with a north wind blowing and it's dark out not much after four? Some projects Nelson could work on at his seat by the woodstove, but he spent many of those winter evenings writing letters by the light of a kerosene lamp.

Madge remembers that her father was quite a writer. "I work for Home Support, and Robin Monrufet was the manager. The Monrufets were the ones who had the *Lady Rose*. My father had written George Monrufet, the father, back in the 50s. But his son Joe kept the letter all this time, and he gave it to me. Then somebody else that my father had known from Kildonan, he'd written her a letter and her daughter gave me that one." She was amazed at how many people kept her father's letters for so many years.

My wife and I did the same—we still have all the letters Nelson sent us over the years. They have been especially helpful in writing this book because he usually wrote them over a period of several days. Then Mary would pick them up and mail them for him. So his letters are like mini-journals, a window into his life on the island that week. Others who provided copies of his letters include Pat, Ron, Patty, Heather, Inger, Dave and Leona. I am sure other letters are buried out there somewhere, and if you find one, I would love to see it!

Patty told me, "Nelson always had a fountain pen and a bottle of ink handy, and that is what he wrote with. He would dip his pen and write." Nelson occasionally wrote in script, in which case he seems to have used a ballpoint pen. But most often, he wrote in his unique style of calligraphy, always with his fountain pen. Debbie told me, "Nelson carved his return address in the heel of a shoe, and that's what he used to stamp his address on the envelopes. *That* should go in the book!" She added, "His letters were done so beautifully."

Heather wrote back and forth with Nelson from the time of her first visit to Copper Island in grade 11, and she remembers his kind and thoughtful responses. Early on, when she was looking forward to missionary training at Camp Ross but feeling self-conscious about wearing clunky gumboots in the wet climate, Nelson wrote back, "Now about the boots, don't let the boots worry you; it's what is in the boots that matters. I am quite sure that you will look just as sweet in boots as the other girls. It's the YOU that Matters… be sure I will continue to remember to pray for you. He then quoted the Bible (King James Version, of course), "How beautiful are thy feet with shoes (BOOTS), O prince's daughter!"[2] Heather has loved gumboots ever since.

As mentioned, Nelson made Heather a carving that incorporated her favourite flower, bird and Scripture. But Heather was leaving for Ontario, and Nelson knew she would be unable to take the large carving with her. So he mailed her a miniature version (with the addition of a frog and a pond), set in a small wooden box. The larger version she has now given to Madge, but the miniature is a keeper.

The "portable" carving Nelson made for Heather when she went to Ontario for a time and couldn't take the larger version he had given her. This one arrived by mail, packed in a lovely handcrafted box. Photos: Heather Arnott.

Here are a few more tidbits from Nelson's letters:

Thanks for your writing, Sarah, you write a good letter, and you Benjamin, thanks for the picture you made for me, every time i study it i see more detail.[3]

Do you sometimes wonder if God is so busy with the affairs of this universe He forgets you?[4]

i have just been going through a very deep mental depression and at times i wonder if i will lose my mind. Some say… xxxx. Well now.[5]

In short—all my girlfriends are getting married and leaving me stranded high and dry on the beach. Poor miserable of bachelor that i am. So many people are considerate and kind to-me-ward but at the same time i have so great a feeling of being alone, no one to love and confide in… It is such that i do not feel a part of the human family.[6]

Monday night. Here in the Hermitage with a fire in the wee stove and Snuggles sleeping on the bed with one eye open. Even with such good company, this life of a hermit is very lonely at times. There are so many things to talk about, things which a little dog (be she ever so loving) does not understand. It must be so with us and God. Though we had the understanding and wisdom of countless ages and though we could quote every book ever written, yet i trow we would know no more of God's boundless wisdom than Snuggles or yon cat knows of our thoughts. To sum it up, i remember in my book that i had wrote:

How high is up
How deep is down
How far is distance all around.[7]

There are times i get very dejected. i have so many ideas, plans, etc., but no one to give me any encouragement. i want to do so much for the cause of Christ but it seems i am hampered at every move.[8]

Well, things move slowly on Copper Island but i am still here and will be so long as the Lord wishes.[9]

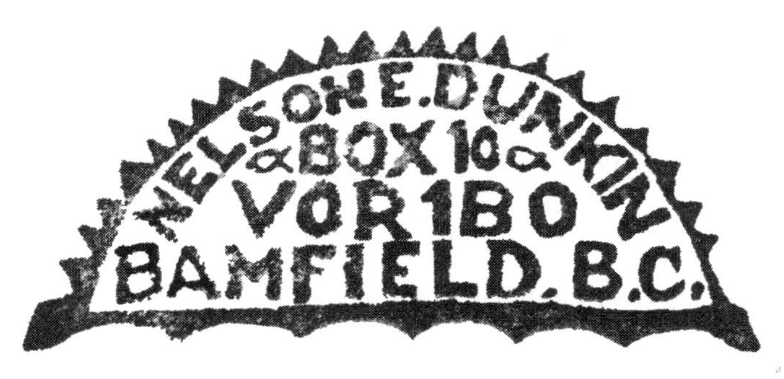

Nelson carved his return address into the heel of a shoe to use as a stamp on his envelopes. Copper Island Camp still uses this mailing address today.

---

[1] Letter to Jim and Sarah Badke, February 14, 1988

[2] Letter to Heather Arnott, January, 1978; Song of Songs 7:1

[3] Letter to Jim and Sarah Badke, February 14, 1988

[4] Letter to Pat Rafuse, May 12, 1982

[5] Letter to Pat Rafuse, February 20, 1983

[6] Letter to Pat Rafuse, December 12, 1982

[7] Letter to Pat Rafuse, December 12, 1982

[8] Letter to Pat Rafuse, undated

[9] Letter to Len and Linda Gedak, undated

Be Thou My Vision

Be Thou my Vision, O Lord of my heart;
Nought be all else to me, save that Thou art—
Thou my best thought, by day or by night,
Waking or sleeping, Thy presence my light.

Be Thou my Wisdom, and Thou my true Word;
I ever with Thee and Thou with me, Lord;
Thou my great Father, I Thy true son;
Thou in me dwelling, and I with Thee one.

Riches I heed not, nor man's empty praise,
Thou mine inheritance, now and always;
Thou and Thou only, first in my heart,
High King of heaven, my Treasure Thou art.

High King of heaven, my victory won,
May I reach heaven's joys, O bright heaven's Sun!
Heart of my own heart, what ever befall,
Still be my Vision, O Ruler of all. A-MEN

Nelson's favourite hymn was "Be Thou My Vision." He loved the words, which describe well his beliefs and values in life, but also its traditional Irish tune. The hymn is a translation of a sixth-century poem by Dallán Forgaill, set to the Irish folk melody "Slane." Very popular in the UK, so perhaps also Mina's favourite hymn.

43. Poetry and Prose

i never cease to marvel at the wonders of reproduction: Why, even the trees have baby trees.[1]

In addition to letters, Nelson enjoyed writing poetry and his random thoughts. Sometimes it was difficult to know if the writings he sent were original to him or if he had found them somewhere. So I have included the poems and sayings to which I can find no reference anywhere online, assuming them to be his own creations:

How people are driven by Satan crazy over money, so it is a cause of higher, higher and higher. Some few years ago i wrote a poem on the way the world is going and on money i wrote:

> *Greed for money is a sin, 1 Tim 6:10*
> *Money buys the souls of men. Mark 8:36*
> *Money abounds as flakes of snow, Luke 12:19-20*
> *And prices spiraling upward go. Rev 6:6*
> *But when the Son will frown thereon, Matt 13:33*
> *Like sun on snow 'twill all be gone. James 5:1,2,3*[2]

Dears:
Keep well, keep happy, keep the faith
These horrid last days of this wicked old world.
Love & Prayers, Nelson[3]

Guide me Dear Lord-
Please show me the way,
For left to myself
I may falter and stray.

Lead me Dear Lord,
For I never have been
This way before
So beset by sin.

Hold me Dear Lord-
The way is so steep,
The mountains so high,
The chasams so deep..

Show me Dear Lord
How to be humble,
For without thy hand
I'd surely stumble.

Protect me Dear Lord-
The night is so cold,
The wind is so bitter,
The wolves are so bold...

Thy voice Dear Lord
I pray to hear,
For it dispels
My faltering fear.

Teach me Dear Lord,
That I might be
In all my ways
More like unto Thee.

Copper Island
May 9th 1978

4

Another poem, which I won't include here as it is six pages long, was titled "Foolishness Extraordinary" and explored the imagined backstory of a familiar nursery rhyme. Here is a sampling:

"There once was an old woman who lived in a shoe
She had so many children she didn't know what to do.
So she…??"

Three hundred years ago
(from the toothmarks this we know)
Some illiterate mouse gnawed away
What the last line had to say.
And to this day in history
The words that mouse ate are a mystery.

Confusion has the rule of the day
But i shall that old dragon slay.
With my many degrees and superior mind
i launch forward the facts to find.
And with my Government grant i cannot fail
To solve the mystery of this tale.[5]

The rest of the poem is a creative but rather sordid tale about a godless scoundrel who married someone against her will and gave her an enormous old shoe in which to raise the family. He eventually turns the shoe into a boat and sails away, but it founders in a storm.

---

[1] Written on the side of a poem, sent to Jim and Sarah Badke, undated

[2] Letter to Jim and Sarah Badke, February 14, 1988

[3] Written in Christmas card to Don and Patty Cameron, undated

[4] Sent to Heather Arnott, May 9, 1978

[5] Sent to Jim and Sarah Badke, undated

44. The Solitary Life

"Now there is great gain in godliness with contentment, for we brought nothing into the world, and we cannot take anything out of the world. But if we have food and clothing, with these we will be content." – the Apostle Paul[1]

With Mina gone, Nelson settled into a lifestyle of being happy with the very basics of life. A bowl of Irish stew, a banana and a reliable set of suspenders, and he was a happy man. Happier still if someone came by with a chocolate cake, of course. Nelson subsisted with the essentials but welcomed a feast if it came his way, and he welcomed the people who brought it even more.

What is Irish stew, you might ask? Typically, red meat with carrots and potatoes, but Nelson wasn't picky. "He'd make us dinner," Leona remembers from times of visiting her grandfather, "which was like some sort of hash. Just with meat and everything in one pan." Most of these ingredients would come from cans since Nelson had no refrigerator. Bill Priest was impressed with his cooler, though. "In the middle of his house, he had a shaft from the ground up through the roof, boxed in, and the cool air drafting up from the ground was always cooler than the air inside the house. There was a door to it, like a cabinet door. The groceries that needed to be cool went in there on wire racks."

During the couple of years that Don and Patty lived with Nelson, he fared much better. "Nelson had no teeth, but you would never know it," Patty told me. "He chewed everything with his gums, even steak. Nuts were the only thing he didn't eat because they were too hard. Whenever I called for a meal, I'd say, 'Lunch is ready,' and he would drop everything instantly and come. He never wanted to keep me waiting for a minute. And if anybody needed any help, let's say Don needed help with something, he would always put down his tools and go and help, and then go back to his work. He was always that way. Very considerate. He lived a simple life and he was very humble."

Patty also remembers the pot of comfrey tea that Nelson kept brewing on the stove 24/7. He had comfrey plants growing out behind the house—it tends to take over like a weed—and he constantly added more water and leaves to the pot. "He poured off the pot into his cup and drank that tea all day. It was very healing for him. And good. Yeah, he really liked it." Nelson had a reliable source of water from a small spring just above the house—clear and pure, freezing cold and faithful even when the creeks on the island dried up in the summer.

Certain foods Nelson could never get enough of, especially bananas—which he called "monkey fruit"—and chocolate. Pat remembers, "Tom and Debbie made a chocolate cake for his birthday and brought it, and I went with them. Debbie unloaded everything from the boat and set it down on the dock, including the cake. Well, along comes Nook, right? So Nook helped himself to the cake. That was the end of that. My mom made Nelson another chocolate cake, and they took *Lady Rose* and they stopped at Nelson's on their way to Bamfield. They made sure to get off the boat and take the cake to him."

Brian remembers how much "Nelson liked it when somebody else came and cooked for him." Pat added, "I would get the woodstove going, and I would continually keep on a pot of hot water, especially in the new house with no running hot water. You had to bring in the water from the creek.

Nelson loved to wash the dishes. I remember that Nelson could really stack up the dish rack, and he would say, 'No, you can't dry the dishes. God is drying the dishes.' Don't dry them. All the dishes could wait. That's a good memory."

Bill Priest said, "I would try to get over a few times during the year to bring him dinner and have it with him. I would bring over a prepared meal with cut-up vegetables and so on. And it took me a while to realize he didn't have any teeth, so he would always leave the vegetables. When I found that out, I asked him if he would be interested in going to the Barkley Sound Resort to have dinner with me and Jim. I would bring him back at the end of our evening. I was a bit surprised that he would actually choose to leave the place. But it was never for overnight; I always brought him back at the end of the evening and would head off home."

Groceries were the reason Rich first met Nelson. "Friends of mine knew this man who was very different, living on Copper Island. I had a boat, living here in Ucluelet, and Copper Island was an hour and a half by water. The dear folks at Tofino Bible Fellowship and the House Church, of which I was a pastor, were very concerned about Nelson Dunkin. So they took an offering, and I bought a boat-full of groceries with the enormous sum (for those days) of $55.

"I found his place just around Clifton Point. His float was a large and random raft, and you had to play log roller across to his place. He was waiting for me and was obviously very glad to see me. His wife Mina was still alive then. Jonathan and I took a couple of boxes of groceries up to his house. Unbelievable. It was four storeys high. An amazing place. And he said, 'Good thing you brought groceries because I was about to start chewing the bark off trees."

Nelson's granddaughter Leona said, "I don't think he ever washed his sheets. And he slept with his dog all the time. I remember the side room off of the kitchen. He had the same sheets all the time. And Snuggles and him would sleep in the bed together. It was a hairy mess, and he didn't care. There wasn't much effort at cleanliness. He'd always wear those pants that were wool, so he didn't ever have to wash them. And he had long fingernails. But I loved his dry sense of humour."

Patty said the same: "He wore his same clothes every day, with suspenders. He hung them up at night and put them back on again the next day. We always thought he must only have a couple of shirts, until one Christmas when we gave him new flannel ones. He opens up the parcel and says, 'That's great. Isn't this really nice of you?' And he opens up a cupboard, which was stacked high with new shirts, and he puts them in and shuts the door. I mean, what he was wearing wasn't worn out yet. He did have a brand new shirt he always wore every Sunday."

Debbie recalled, "One time in the spring, we went and visited him and took all his dirty laundry—sheets and everything—to a laundromat in Port Alberni. He would put a sheet on the bed, and the next week, instead of changing it, he would put another sheet over top. We had to wash things about three times, but it was nice to take it all back home, and everything was fresh."

More and more, Nelson felt attached to the island and was reluctant to leave it for any reason. Madge told me, "To say that my father was afraid to leave the Island for fear of what might happen to it is an understatement. I sometimes think he did have post-traumatic stress disorder. People didn't talk about things like that in those days. I know he was quite depressed at times." From my conversations with him, Nelson carried a great sense of responsibility for "God's Property" on Copper Island, and its future weighed heavily on his mind.

---

[1] 1 Timothy 6:6-8, ESV

45. The Camp Dream

Yes, i sure wish to see a colony of Christians here on God's 105 acres. Seems everyone could do so much to help the others. It is not a small responsibility to hold the Title to God's Property. One needs to be very cautious just which ones to invite, be very cautious of getting any moochers and deadbeats or those of an executive trend of mind.

To date this is my list:
- Jim & Sarah Badke & Children
- Don & Patty Cameron & Children (Troller)
- Bernd & Sylvia & Johanna (Troller)
- Neil, Kathy & Joshua Harmsworth

These are all Christians of the highest order and hard workers.[1]

Nelson believed that a small community of people living and working together on Copper Island would be an ideal situation that would greatly benefit its participants and further the kingdom of God on earth. Knowing his background, I wonder how he reached this conclusion. I also wonder to what extent his dreams influenced my own vocational focus. We once owned a large, cedar-clad house that looked like (and was even older than) Nelson and Mina's. For 15 years, my wife and I led gap-year programs for young adults, drawing them into a small, temporary community for biblical studies, adventure and service in the world. We loved the years that we shared a big house with groups of students, living and working together. We indeed saw some wonderful things happen, though the experience was not without its heartbreaks.

Nelson wrote to Pat, "Now about all those young people out of work and fiddling away their time in the city. If Christians would only put aside selfishness and cooperate, they could all go in for a farm. The ones who could get a job could supply money, those out of work could be working on the farm. Everyone should be married and happy because they are doing something for others. Those on the farm could supply produce to the city workers at a great savings. Those at the farm could have wonderful Spirit-filled meetings every day. Crazy world—we go to the city to buy our food—where do the big stores get their food from? Is it created in the city?"[2]

Nelson II told me, "My dad was happy that the property became a camp; this delighted him. But I don't think it was his original or final dream. I think his main idea was that it would become a permanent residence for Christian people. I think that's what he was aiming at. And he spent years bemoaning the fact that people would only come and visit, but nobody would ever come and homestead there with him. One of the last times I went down to visit him, he was well into his 70s and he could hardly walk. He was stooped way over but he hiked down to the waterfall, and he was down there digging and working and preparing to get a waterworks going for the Toy Shop. I think that was his idea—to have a group of people there."

As the years went by, the dream of a community on Copper Island shifted and refocused. Nelson came to realize that he would not be the person who would bring this about, though he always hoped to stay involved. Even after illness forced him off the island, he always expected to return there one day and be part of the island community. He also felt the weight of the responsibility of handing off his dream to someone else. He wrote to Pat, "Were it not that i am still holding title to God's Copper Island Property—it would be quite likely that i would take off to regions unknown.

"It is very true that Copper Island could be deeded over to some Christian Cause but of such i would need to be very cautious indeed. Extremely cautious. Would they develop the place to the Glory of God or would some money-hungry Board of Directors sell the property to get new cars, fancy houses and money to pay the interest on unwise loans from the Money Mongers?

"Nelson II says he would like to bring camps here but apparently he has no followers with any inclination at all to come here and do any much-needed work."[3]

Bill Irving remembers, "We often chatted about what he thought was the future of the property. He said, 'Well, what I am thinking is that Pebble Beach is where I should have built originally.' We walked over to the beach and had a look. There was the creek at the end and the flat area above the beach. Nelson said, 'I hope someday that this is part of a camp.' I think he sensed this was going to be a permanent fixture on the coast. But to appreciate that he and Mina had always thought and prayed about that, and it has come to fruition the way they hoped—that is a good testimony to their faithfulness." Today, there are two clusters of summer camp cabins on that very spot above Pebble Beach.

Camps and retreats had already taken place on Copper Island, almost from the time when the family first moved there. Bill Irving continued, "The times we went in the summer with kids were an eye-opener for those young people. Just to come and see a person living like that and all the creative things he did without power. To experience the quiet. And the Christians were not singing hymns and walking piously around all the time. But they were having fun, and they were creative, and they laughed and sang. The experience made a huge difference for these young people.

"The kids would fish in the harbour, and we'd get a piece of plywood and tie it to the back of a boat because we couldn't afford skis or anything. And the kids would ride on this plywood and fly all over the place. It was one of those places where you could adventure like that without somebody waving their finger at you. That was my sense of the young people coming and experiencing the quiet and the alternate lifestyle. They didn't have to succumb to all the pressure they were accustomed to in schools and cities."

But there were definite liabilities and risks to running camps on Copper Island, including limited access, buildings not built to code and the danger of fire and storm. Roy discussed the reasons that Coastal Missions didn't take up Nelson's offer to run camps and have their base there. "He always wanted to have something to do with us having camps on the island. But the liabilities with his wooden buildings were not worth it; we couldn't legally do it. I don't think he understood the way the world works these days."

In 1987, Nelson wrote to us, "Getting sleepy—praying much for a leading as to how to will this God's Property in such a way and to such a person or persons so those coming to live here would always be sure of their tenure and also so the property and timber would be free from exploitation, and also such that no fake cult would find a footing here, and to be as free as possible from government interference."[4] His prayers for suitable partners would soon be answered, though it took longer to discern the right fit for Copper Island.

Unfortunately, wolves in sheep's clothing also had their eye on Nelson's property. Joan said, "Some people in boats were causing trouble, trying to weasel coastal people out of their properties. Some guys came in and tried to convince Nelson to sign a contract." Tom added, "Then Angie, from the village across the way in Sarita, found out about this and let Mary know. Mary went straight out there and tore up the contract." Dave from the Coast Guard also became involved, as he was very protective of Nelson's interests and was especially concerned because some of these scoundrels were coming up from the US.

The first reputable organization to show interest was the Canadian Sunday School Mission, now known as One Hope Canada,[5] a thriving ministry that runs dozens of Christian camps across the nation. Bill Irving said, "We had gone down with the Canadian Sunday School Mission because Nelson had talked about getting rid of the place. So I brought them over in our boat, and we had a pretty good chat because Nelson wanted it to be dedicated to kids' work. We left with some optimism that here was a big organization that knew how to run these things. But the next day, we found out that Mary heard we had been there and raced down to talk with Nelson." He decided not to go with this organization.

In the end, it was a connection in the State of Washington that looked like a promising fit for Nelson. Mary had helped with some camps in the Cowichan Valley run by the Pentecostal church. She met a group of people from the Philadelphia Church in Seattle who had purchased a tug called *Covenant* to use as a mission boat on the coast. Mary told Nelson that this group would be interested in running camps at Copper Island. Taking her advice, Nelson invited them to come. I talked with Dave and Inger Logelin on the phone, and they sent me this account of what happened:

Inger: "We first met Nelson Dunkin in 1987, when the first camp was held at Clifton Point, Tzartus Island, at Nelson's invitation. He would have been in his late seventies. We arrived by tugboat with a group of workers and campers, organized by Sharon and Jeff Watkins and Earl Cathers. The first 43 campers came from the reserves at Saanichton, Nanaimo, Ucluelet, Bamfield and Port Alberni, with workers from the Vancouver, BC and Seattle, WA areas. I served as counsellor to the senior high girls. Nelson would take his meals with the campers, sitting on a log and thoroughly enjoying a hot meal and a bit of company, then disappear into his two-storey house on the water.

"Nelson told us that his vision was to have the property used by God for a Christian camp and retreat base. He had the following letter read to the workers at the first camp:

To all Campers, Staff and Ministers of
Copper Island Wilderness Camp:
GREETINGS IN JESUS:
As i am not fluent of speech Roberta has kindly offered to convey my sentiments to you.
Please be assured that i am greatly im=pressed with your good behaviour and friendly=ness.
Also i would i had the ability to convey to you, each and every one, how much your efforts and presence here means to me.
I have had times of discouragment and bitter dis=apointments in my desire for God to use this beautiful place to His Glory and now at last, after many prayers and the prayers of dear friends, my vision is miraculusly coming to pass before our very eyes and you dear ones are a very necessary part of that Vision.
For so long as you strive to please God this camp is as much your's as mine.
Now until we meet again may our Lord be your portion and blessing
Servant of His Servants
Nelson E. Dunkin

"He said he had prayed for thirty years for this to come to pass. Nelson Dunkin welcomed us to use the 105-acre property with its old-growth forest, deep sheltered cove, mile of beachfront and water source from its picturesque waterfall for a Christian camp with just a 'gentleman's agreement.'"

Dave: "In July of 1987, a friend of mine flew me up to Copper Island, where my wife was volunteering in the first Native Bible camp. I was there for only a day, and that's when I met Nelson Dunkin for the first time. While we visited that day, I was impressed that Nelson was a very humble man who cared for people. He was strong in his faith and said that day that he and his wife Mina wanted the property to be used for God's purposes."

In August of 1987, Nelson wrote to us, "We had a Wilderness Camp here. The good ship *Covenant* came from Seattle with 12 all-eager workers and tons of equipment and food. They got the big two-oven range all shined up like new. Shelves and tables, a sink and water hosed over from the water line. The brush and ferns all cleared away, and rough benches, and over all a huge 40' X 50' tarp. The trail you surveyed to

Pebble Beach was chopped and hacked out, and down at the waterfall was a shower-bath with hot water.

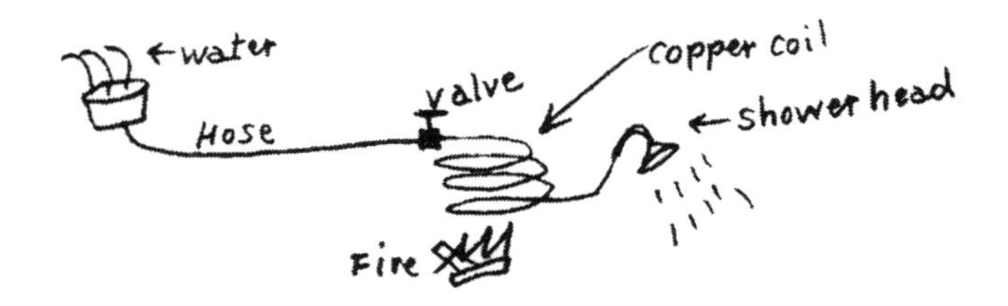

"The campers had tents all around in the bushes along the beach. 38 campers, age 12 to 19, + 3 children.

- 19 Willing Workers
- 20 Saved
- 23 Water Baptized
- 8 H.S. Baptized
- 1 Refilled

"i am still in that state of wait-wait-wait. It has been suggested to put this place as a Foundation. And we are looking for, not a Christian Lawyer, but for a lawyer who is a Christian. So far my son has not managed to get out here so we continue in prayer and waiting."

Nelson wrote to us in February of the next year, "This summer, they plan to hold three Bible camps here. As yet there is nothing done on the REDTAPE of it yet… You see why i am very cautious with the handling of God's Property to do everything possible to keep it out of the clutches of the devil and the heavy hand of government. i imagine these Seattle people, being building contractors, will make everything modern and city-like."[6]

In May, Nelson wrote: "Saturday. Mary came with your letter and i am answering right away so as to give you the dates of the Bible camps. They are having 3 camps this summer:

- July 11-15 - Junior 9-11
- July 18-22 - Junior Hi 12-14
- July 25-29 - Senior Hi 15-19

"In June they expect to land here from Seattle with the *Covenant* towing a 2-engined Landing Barge carrying a No. 40 Wood-Mizer sawmill and, if they can get it, a John Deer Bulldozer, with a volunteer work crew. i am sure there will be some very interesting people in the work crew. There is no knowing how many campers there will be.

"As yet there have been no agreements made on this God's Place. i want to be fair in every way but at the same time have definite 'Thou shalt not's' in the agreement for as much their protection as ours. i guess i have been seeing and reading too much how Satan uses people, innocently or otherwise (Judas for example) to try to wreck God's plans. So it behooves us to try to keep in step with Jesus."[7]

Donations continued to come in toward the needed facilities. Dave wrote to Nelson, "When I read the letter World Vision sent about their donation of the sawmill, I was really happy to see how God answered that specific prayer. There have been many things happening this winter, big and small, giving us a real assurance that the Lord really has his hand on the building of the camp. I'm sure you have felt discouragement sometimes, as we all have, but the Lord has his hand upon you. We have been praying for you and the camp project and are looking forward to this summer, working with you and the others that God will send to help. We want you to be assured that this is your home and we will just be working together with you to fulfill the vision for the camp that God has given you and that you have kept through the years."[8]

Nelson replied to Dave and Inger, "A rip-snorting southeaster today. So good that it held off for Mary to come yesterday with your letter. The Lord has drifted in a wonderful log, clear and straight, and I think it will be close dark-grained like mahogany. Now to get my Kildonan friends to saw it when they can. So soon as you can, keep Mary informed of your camp schedule so she can plan her vacation Bible school. I think you are going to have difficulties as one little guy says he's coming to all three camps, regardless of age. I should be doing so much, but with this wet cold weather I do so little, and that mostly shop work towards my toy factory."[9]

Dave and Inger came to the island to stay full-time in June of 1988. "We began building the camp and a few buildings, with his permission and blessing. Nelson was there for a short time that summer, but was not well. We had a team of people constructing our first building, and working on the sawmill cutting lumber. He rarely came out of his house, but one day he came out and stood looking at all the work that was going on. With a big smile on his face, he said, 'They are all as busy as beavers.'"

The Bible Camps at Copper Island were a great success - Praising the Lord.[10]

---

[1] Letter to Jim and Sarah Badke, October 25, 1987

[2] Letter to Pat Refuse, May 12, 1982

[3] Letter to Pat Refuse, December 12, 1982

[4] Letter to Jim and Sarah Badke, April 18, 1987

[5] One Hope Canada, onehopecanada.ca

[6] Letter to Jim and Sarah Badke, February 14, 1988

[7] Letter to Jim and Sarah Badke, May 2, 1988

[8] Letter from Dave Logelin to Nelson, February 6, 1988

[9] Letter to Dave and Inger Logelin, February 14, 1988

[10] Letter to Ron Pollock, undated

46. Getting Sick

All of Nelson's dreams and plans had him living happily on Copper Island for the rest of his life. He never expected illness to make this impossible. Indeed, he would not have wished that scenario on anyone. Years before he became sick, Nelson wrote, "Mary is quite upset that the Happynooks might be moving to Victoria. Brother Happynook is failing and much worried over the prospect of going away as he wants to stay where it is Home to him. i too think it better for him to stay put. After all, is there not a little hospital and nurse in Banfield? If you make a study of it i think you will find that far more people depart this life from Hospitals than otherwise. The thing that really matters is not *from* where they depart but *to* where they depart.[1]

Perhaps it was the damage to his lungs during the war that made Nelson susceptible to pneumonia. Nelson II remembers, "My father had a doctor who used to come from Washington State every year on his yacht to see him. How he came in contact with him was that my father got pneumonia, and he prayed that a doctor would come by. And this doctor sailed into the bay within a day or so and got to know him. The doctor treated him and gave him a prescription for penicillin. My father recovered, and they became very fast friends."

But the pneumonia would show up again. In May of 1988, Nelson wrote, "Thanks to Mary, yesterday one of the Coast Guard's sausage boats, four of a crew and the nurse from the Banfield Hospital arrived—it seems i have new-moan-i-a, so she left me some pills to take. Some years ago, the nurse in

Banfield said i didn't have xx because i didn't have a fever, but after having it for a year, Dr. said it was."

And it returned once more. Bill Irving told me, "As Mary was getting older, she couldn't travel as much, and Nelson counted on people stopping by to check on him. We were lucky to come by on that weekend." He had found Nelson very ill. "He was in the little cottage down near the head of the dock, bedridden with pneumonia. After we arrived, we were part of the arrangements to get him out of there and to the hospital. Mary Scholey came from Bamfield. That was the last time we saw him on Copper Island. He was in very poor health and couldn't feed himself or take care of himself."

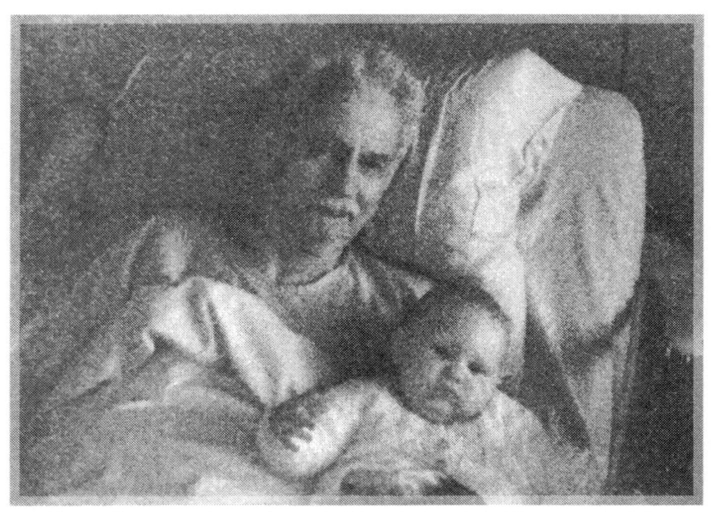

Rare and grainy Polaroid photo of Nelson in hospital with someone's child.

Nelson II remembers, "My father was rescued off the island and brought to the hospital in Port Alberni by the Coast Guard in Bamfield." In June of 1988, Nelson wrote, "Got your ever welcome letter and feel bad that i can't be the help to you i should. Here i am in the Port Alberni hospital now 3 weeks (Madge says 5 weeks) with no idea when i will be out or in what condition i will be in. Mary, God bless her, is ever a great help. With all the work she does i don't like to burden her but after all there is no other to turn to. Mary is so reliable that i have told her to take charge at Copper Island. That what-so-ever she says is LAW. The Seattle work crew are there now."

Nelson II continues, "He recovered somewhat, stayed with Madge for a while and ended up going back to the island, leaving Snuggles with Madge." Dave explained, "Nelson became ill and was hospitalized in Port Alberni in 1988. In the fall of 1988, he was back on the island for a time, recuperating at home. But in November, he became so sick that he had to return to the hospital in Port Alberni." By now, Nelson was desperately ill. Here he tells his story:

> *An Explanation:*
>
> *Perhaps one reason you have not had word from me for too long a time is that for six months i have been in the hospital. It would seem that the Prince of this world has chosen me out—a kind of 1988 Job [it was actually already 1989]—but thanks be to God for giving me his strength to endure the testing, which has not been easy. Just now i have no pain but there was days and months when i was in pain continually. For 2 months i was in Intensive Care; out of my mind; knowing nothing. i was told that twice they thought i had departed this life. i was far into that formidable, dark Valley of the Shadow of Death when God in his love and mercy reached in his hand and gently drew me out. All credit to the saints who were praying for me day and night.*
>
> *Now it behooves me to be at doing some kindness to others along this rough and trying way.*
>
> *Lord willing i must return to my duty on the Island—there is so much to do that i need not be idle. i must be ever mindful*

of what Jesus has done for me. The doctor (though not yet a Christian) told No.II that i am a miracle.[2]

Madge told me that her father had been in the hospital in Port Alberni for some time and was not improving. Her brother came from the Mainland and arranged for Nelson to be moved over there. Nelson II said, "They had put him on the end of the list for palliative care. But at that time, my ex said, 'No way, we're going to get him over here.' We had people here who had a good doctor in Richmond who said he would take him. So we made the arrangements for my father to be shipped over to Richmond General Hospital. He was on life support, and he was in a coma for many months. They were able to keep him alive, but he was getting worse." Madge and family came over to say goodbye to her father, and she was surprised that the doctor still had hope of "fixing him up."

But Nelson did recover. Nelson II explained, "We had a fellow come in, a preacher from Australia, whom I asked to pray for my dad. He was leaving for home but he walked into the room just before he caught the plane in Richmond. He laid hands on my father, prayed and walked out. And the next day, the hospital called us and said that my father had come out of the coma and was miraculously on his way to recovery."

While out of this world and unable to speak, i wrote, 'Jesus is coming soon and his reward is with him.'[3]

---

[1] Letter to Pat Rafuse, May 12, 1982

[2] Letter to Ron Pollock, possibly May 1989

[3] Letter to Ron Pollock, possibly May 1989

47. Maillardville

"After he came out of the coma in the Richmond Hospital, my father's recovery was fairly quick," Nelson II said. "Shortly after, the hospital said he needed a place to go because they were kicking him out. Our good friend, Francis Keller, was the landlady of an apartment in the Maillardville area of Coquitlam. She rented the place to him. We got him set up there, and that's how it all happened."

I remember being astounded to hear that Nelson was living in Coquitlam—not only because of the contrast to where he had spent the past 30 years, but because I found myself living just 15 minutes away (depending on traffic). I had taken a position as a youth pastor in a church in Port Coquitlam two years before and had entirely lost track of Nelson in the shuffle of both our lives. I soon drove to Maillardville and found the apartment, sandwiched between Lougheed Highway and the TransCanada freeway, next door to an autobody shop.

The rickety back steps to Nelson's place. His windows are on either side of ladder.

At first, I wasn't sure how to find him. I tried the less-than-encouraging front entrance, my feet sinking a bit on the sagging floor. The apartment number I had been given was merely a guess. I knocked at the door and waited. Hearing no movement inside, I tried again. Nothing. I was about to turn away when I heard something, and a muffled voice said, "Come in." Heart beating, I turned the handle, poked my head in the door and inquired, "Nelson?"

The man himself rolled into view. It took him a moment, and then his face lit up. "Come in, come in! Oh, aren't you a sight for sore eyes!" I sat down on an old couch, and we quickly caught up with one another's lives, though he made sure this was much more about me than about him. He was greatly changed since I had last seen him on Copper Island. He could shuffle along with a walker, but the wheelchair was his main locomotion. He was thin and gaunt, and I saw that he had a catheter bag attached to his chair. But his smile was as broad as ever, and his delight at seeing me was complete.

1

The irony of his situation squeezed my heart. When we first moved to Port Coquitlam, Nelson had written from Copper Island, "Just looked at a map and see that you are right in the midst of the Great Vancouver Rat Race. i do actually feel sorry for you."[2] Peter told me about Nelson's usual term for apartment buildings: "'Human hutches,' says Nelson." But now Nelson was the hamster. For the next several years, I made it my goal to visit him there at least once a month, often bringing my kids to brighten his day—and ours.

Madge told me, "My father never wanted to live in the big city. I don't think that Coquitlam was a good scene for him. The last place he would have wanted to be. I'm pretty sure he probably was discouraged in that place, but that was my brother's doing." Bill Irving remembers some disagreement about where Nelson should get medical treatment and support. "I had a very general conversation with him about why he ended up where he did, and I think it was more to do with his son being the senior person." Madge tried to have her father moved back to Port Alberni, arranging with the doctor there to take him on. But by some miscommunication, Nelson thought the family in Port Alberni didn't want him there.

As the fumes from auto paint wafted in and various less-than-reputable people walked the alleyway below his window, I wondered how someone like Nelson could end up in that crumbling urban apartment. Pat remembers that Nelson called the place "Tin Can Alley." I agreed that it seemed like the last place he would want to be. Yet two things redeemed this setting for me. One was that, except for his appearance, Nelson was unchanged. He laughed, made wry statements about the government and society, showed me his latest projects, and generously offered me whatever he happened to have about him.

Also, as I continued to visit regularly, I soon saw that everyone around him loved him. Surrounded by old suburbia, Maillardville is a small community in its own right. Nelson II talked about the owner of the little Chinese grocery store on the corner. "She and her daughter used to go over and visit him, and he would buy things from her. They loved him. Mary would send money to the store, and the lady would go over and see what he needed and bring it to him. A monkeyfruit a day. Ice cream was also on the menu. And hot chocolate." Nelson II regularly picked up things for his father at the nearby Superstore.

Nelson also looked forward to the arrival of a Meals-On-Wheels volunteer, which gave occasion for an article about him in the local paper:

> *No man is an island, said John Donne, and even living on one for 30 years has not left Nelson Dunkin a solitary man. He lives a solitary life, however, a consequence of an illness that pulled him from his Copper Island paradise to a second-floor apartment in Coquitlam. The building has no elevator, he is in a wheelchair. The landlady shops for him.*
>
> *Although he has letters on his table and drawings from his grandchildren on the wall, he doesn't get too many visitors these days. "Not as many as I should. This is such an isolated place to come to." Sometimes he gets lonely but doesn't complain. "It's not such a bother to be alone when you know somebody's going to come." Five days a week, somebody from Meals on Wheels comes.*
>
> *"And they're very good—everybody is most kind and helpful. They can't stay long although they would like to. They're very short of drivers so they can't tarry long."*
>
> *"I don't want a television, thank you, because we hear enough about crime in the newspaper without looking at the television." But he does have a small radio. "Sometimes I like to hear a good sermon or good music. None of this rock business." The absence of electronic interference resembles his island life, which he said was not secluded. "The* Lady Rose *was going by three times a week and would stop in anytime you wanted to go or come.*

Now, he carves. There is sawdust on the rungs of his wheelchair and tools nearly as old as he lined up on a bench built into the tiny kitchen, the other room in the apartment. "I carve all kinds of funny creatures. I carve angels and give them away. What else could I do with them? People like to be given carvings sometimes." He is now carving a sign for a chapel at Pachena Bay, with the two-word name on either side of a cross. The wood comes from Copper Island. He relies on people to bring wood to him. "I'm like a hog on ice with his tail froze fast. I'm stuck here on account of my condition."[3]

The finished product, now hanging in the dining room of Copper Island Camp.

Ben used to visit Nelson there. "I remember the day he told me, 'I think I want to start carving again.' It must have been a satisfactory thing for him to express his creativity with wood. He also listened to the news on the radio and would say, 'So what do you think about this?' It was a pretty rundown apartment that he had, and seemingly he was out of place there. Down on Copper Island, that's where he belonged. But a time does come in most of our lives when we have to make a move."

Dave and Inger visited him too. "We would bring him wood from the island to carve, and food and little treats. We would often find when we visited that he had given away anything of value that we may have given him. He was always interested in how the camp was going, and I always felt he was pleased. He missed being near the water, and he would comment sadly, 'There are no birds here.' But he enjoyed the eclectic residents of the apartment and having so many visitors."

The brothers Bill and Steve Priest made a point of getting there several times. "After leaving Barkley Sound, I lived in Surrey for a brief time," said Steve. "I was able to visit Nelson in his apartment on Brunette Ave. The visits were always delightful and positive." Bill remembers, "I was back and forth to Vancouver a few times, and would make an effort to stop in and see him. A very simple—and kind of horrible—place, compared to where he used to live. I'd come in through the back, up those rickety stairs. I think the first time I came to the door, somebody was there with him finishing up some care work, and he let me in. And I had the rest of the time with Nelson to myself. It was sad in a way, so I just sat with him and told him what was going on out there for about an hour or so and moved on."

Nelson II told me, "It was just a bachelor suite. I brought a bed in from our place to get him in there. And later, I did the negotiations to get him a hospital bed. He never locked the door because he wanted it that way. He was there for several years and had tons of visitors. People from all over used to come and search him out and go there. But I have a feeling there were a few people who padded their pockets with his generosity. He used to feel sorry for some of these young women who would hang around and were having problems."

Bill Irving remembers his last visit with Nelson. "He was in that flophouse in Coquitlam. He gave us a carving there of an eagle with, 'He will mount up on wings as eagles.' The windows were broken and he had rags stuffed in them, and his wheelchair was too wide to get through the door to get out, making him housebound. He was always interested in getting wood for carving.

"And he said, 'Yeah, the girls next door do all my shopping and check on me every day.' That was a pretty interesting sequel to living as a pioneer on the West Coast, a recluse—to move into that environment that was totally foreign, but to still be gracious and accepting of people, whoever they were, in whatever position they were in their lives and to be thankful for them. I think that message probably still resonates with those girls, an old fellow who wasn't judgmental."

Nelson kept writing letters, especially to those who were unable to visit. Here is a sampling from the time he lived in Maillardville:

> *i try to keep as busy as my equipment will allow me. The 1" X 6" tropical carving board Wynn bought for me: SUNRISE COVE LODGE in 4" letters to go on the new building at Copper Island, which Mary named. Also a small plaque to go on one of the doors. God sure answers some prayers in a hurry. So i had finished the sign and was laying on my bed when someone walked in and here it was Jennie. So she took it to Abbotsford as going to camp she might not have time to stop*

for it. A few days later i had finished the plaque too and was sitting with my back to the open door—wondering how ever i would get it to Copper Island—when someone walked in and stood beside me and sure if it wasn't Jennie on her way to cook at Copper Island, so went the plaque.[4]

Of late I have been becalmed in the Doldrums, in a heavy fog with both anchors down and my sails just hanging, sick-like. Sometimes I wonder will I hold out until the Rapture? Every time I finish one job I say 'Praise God!' and start on another. But of late I have not been so eager beaver. With much grumbling I did force myself to whittle out a little wooden valentine for Mary (about the 30th one I think over the years).

I fear sometimes that I am an ungrateful wretch unworthy of God's blessings. I can't get enthused in gold sidewalks and crowns of glittering jewels. Sweet music, yes, if played on the Ullein Pipes [Irish bagpipes]. My idea is a plain stone cottage along the shore; a level garden plot with tasty vegetables growing; a little stream tumbling down the mountainside and turning a small overshot waterwheel. My precious wife calling from the doorway, 'Come dear, these children have brought their broken toys for you to fix.' So I come in, give them each a kiss and a hug. 'Praise the Lord, have a cookie and a glass of milk whilst I see what I can do with those toys.'"[5]

24th December 1990 From the city of sin and Greed.
My Dear Leona: Greeting you in the Lord

Here it is a sifting of snow on the ground and i write you. i hope you will be enjoying a Blessed Christmas time. Firstly i must thank you for all the gifts from you and the family and especially for that big book on animals. You should not spend so much money on me but be sure i do appreciate your love & kindness.

47. Maillardville

Yesterday afternoon Nelson II & Family were here with all of their little gifts, and towards evening Jenny, the Head Cook and Crystal; Donna the Second Camp Cook and Marla; and Crystal's Father were here. They brought along a little bit of the top of a cedar tree and fixed it all up in decorations. There is no Meals on Wheels today but one of the men came with a nice new board for me and 6 buns which his wife had baked, so very good of them.

Later in the day: Spanish minister, his wife and two daughters just in to visit me. We don't do very much talking as i can't speak Spanish and his wife and daughters can't speak English but we get along just fine. This is the minister i made a plaque for in Spanish.

Better get this note away to you before moss starts growing on it. So Love & God's Blessings to all. Grandpa

Christmas Day and wishing you all the Peace and the real meaning of Christ Mass. Very quiet here, Landlady went to visit a daughter. So no interference, not even a mouse, so i was able to start another carving (all lettering).

Sunday and snowing over 12 inches and still coming down. Now those queer creatures can stop singing about a white Christmas and go out and roll in the white stuff. Too cold for me. i guess i still remember the night i could have froze only not, because of God's mercy.[6]

Of late i have been doing some pieces of carving. Before i start i ask the Lord to spare me to finish the job and so far he has. i have word from Mary that she has a carving project but as yet no details. So i shall pray again to be spared until the job is done. i am always looking for company and company is most welcomed.[7]

But Oh how i long to be working for the Lord making Scrip= ture plaques. I have joyous news! Mary has asked me to make plaques for her windows. And i have written to tell her there is no one i would RATHER make plaques for. She says they attract many to the way of Salvation.[8]

---

[1] Katie Poole, "An Island of Solitude," *Tri-City News*, November 17, 1991

[2] Letter to Jim and Sarah Badke, February 14, 1988

[3] Katie Poole, "An Island of Solitude," *Tri-City News*, November 17, 1991

[4] Letter to Ron Pollock, July 3, 1990

[5] Letter to Jenny Jackson, Copper Island Camp cook, undated

[6] Letter to Leona Dolling, December 24, 1990

[7] Letter to Leona Dolling, October 2, 1994

[8] Letter to Heather Arnott, undated. Photo: Heather Arnott. Mary's window by the Post Office.

48. The Loonie

December 1990, Night of the 27th. This day was warm with a bright sun so very much like summer, only the trees are bare of leaves. Well, tonight i feel free as a bird as i have resigned as Steward of Copper Island and turned God's property over to very capable Saints of the Lord. To teach children the ways of Jesus.[1]

Nelson wrote this in a letter to Leona as a passing comment among remarks about recent marriages and the weather.

For three years, Dave and Inger had run summer camps and retreats at Copper Island on a handshake agreement with Nelson. As Nelson came to realize that his stay on the Mainland was indefinite and his strength was failing, he decided it was time to take action and hand the property over. Even so, he assumed that one day he would return to join the life and work there. Nelson II said, "Even when he lived in Coquitlam, my father planned to go back to Copper Island. He was always talking about it."

Inger and Dave explained, "Nelson saw that we were following through and were committed to continuing camps on the property. He wrote a letter to Mary, saying he wanted to sell us the property for one Canadian loonie. Mary provided the loonie, and we provided the "Yes!" We did not accept the property personally but formed a registered Canadian society. On December 27, 1990, Nelson Dunkin Sr. transferred his 105 acres on Tzartus Island to the three-day-old Wilderness Retreat Society. It was a celebration day for us: for Nelson, to be assured that his vision for a Christian island retreat that would minister love in Christ's name would be fulfilled; for us, as we saw the hand of God do an impossible thing." Though Dave and Inger are no longer involved, the property is still in the capable hands of the Wilderness Retreat Society today.

Dave and Inger were careful to ensure, through independent legal advice, that Nelson understood what he was doing and that the gift of the property to the new Society was his free choice and under no coercion. An independent Vancouver solicitor advised him of the effects of a Freehold Transfer so he understood the "nature and consequences of the transaction and the rights of the Wilderness Retreat Society in connection to it." He sent a Certificate of Independent Legal Advice to the solicitors who were drawing up the incorporation papers of the new society. Part of the certificate said, "I satisfied myself that my client would sign the said Form A Freehold Transfer of his own free will and not under any undue influence exercised by the Wilderness Retreat Society, any of its directors, or otherwise." Nelson did sign, and the transfer was completed.

It was reasonable that questions would be raised by Nelson's family. Nelson II was involved in the proceedings, and for his part, was happy with his father's decision. "My father was a giver, and he had one notion that Copper Island was God's property and he was a steward of it, and that included any money or anything else he had, anything he owned. I thought this was great—it was his dream, and it was good." However, there was a lack of communication and consultation with Madge and her family.

Leona remembers feeling hurt that Nelson wrote to her in passing about the transfer, leaving it to Leona to tell her mom. "Mom was very upset, of course. I wrote to my grandfather and told him exactly how my mom felt and how hurt she was, not even being told by him. He wrote a letter back to me, but he didn't really address the issue. But then, I believe he felt it wasn't even his property to keep or give. I appreciate that now. But at that time, no, I did not. I was more upset for my mom."

Madge's family sought legal advice and expressed concerns about Nelson's state of mind, the transfer and the proposed use of the property. But it quickly became clear that Nelson had acted coherently and independently, and that the property would be used by the Wilderness Retreat Society in the very manner Nelson desired. It was a great disappointment to the family, who had hopes of cabins and logging income. Madge still feels it could have been handled differently. "In my mind, you just don't do stuff like that. But anyway, I've come to leave that and let it go." Leona sees this in her mom: "She's gotten over it, I would say. You have to move on with stuff. Bitterness only hurts the person who's bitter. It wasn't mine anyway. I was just really upset for my mom at the time."

Ben said, "It would be a tough decision when he decided to sell the property for $1. But he had a burden for the native children, and he wanted the camp to be a place where the children could be reached. And that's being fulfilled to this day. I really like Aaron and Julie [the current Director at Copper Island Camp]. They're good people—they and their sweet girls. So the vision is actualized."

In a letter to Mary, Nelson wrote, "Well, I signed my name, Nelson Edward Dunkin and I am no longer Steward of God's Copper Island Property. And anytime the Lord wishes to take me, things are in order. I no longer wish to be buried alongside August Hammond [a Catholic priest purportedly buried on Copper Island] as my carcass would be repulsive to the campers. Just take me out to Cape Beale and dump me into the deep. That was Percy Wills' wish, to be buried at sea off Cape Beale, but something went amiss with his wishes."[2]

In the old apartment in Maillardville, I listened to Nelson tell the story about the transfer of the property. He was not in high spirits that day. "Got a problem with my plumbing," he told me as he adjusted his catheter bag. I thought, surely no one was more affected by his decision to relinquish his stewardship of "God's 105 Acres" than Nelson himself. Other people—not Nelson—would construct the lodge, the chapel and the dining hall. They would be the ones to create the waterwheel at the creek by Pebble Beach and build a more reliable water system. Nelson would not open his Toy Works and—despite his hopes—would never again join the community of like-minded people at Copper Island.

He finished his story and looked at me with that bemused smile on his face. I asked, "So, what did you do with the loonie?" His smile broadened, and he undid the top button of his shirt. Reaching in, Nelson pulled up a string. Dangling on it—by a hole he had drilled—was the loonie. I couldn't help joining him in his chuckle.

[1] Letter to Leona Dolling, December 27, 1990

[2] Letter to Mary Scholey, 1990

49. From Glory to Glory

After a few falls and increasing ailments, it was becoming increasingly apparent that Nelson could no longer fend for himself, and he was placed in a full-time care facility. He wrote, "For a right time i was in a state of confusion after falling. i was put in Amherst Private Hospital (they do not have an x-ray but they give 20 pills per 24 hours). Some weeks ago when i was still in the Big Royal Columbian Hospital, a little angel posing as a nurse nudged up to my ear and warned me that things will get worse. Aside from the purpose, Amherst is a most beautiful place with the greenest of lawns and all manner of trees and bushes from the Tropicals."[1]

Shortly after this, Nelson II rented a large house in Vancouver, where his father would live in their basement for his final year. Nelson wrote to Leona about this move: "For a while things were in quite a turmoil, moving out with no place to go, but thanks be to God we are now all settled down in our various halls and cubby holes. This is an old three-storey house with three baths, many doors and windows and a garage for three cars and my little carvers shop. A kitchen on the 2nd floor and bedrooms 3rd floor up. No.II has finished off another bedroom in the attic. The carport has an electric door. Counting myself there are seven in this mansion. And i would just as soon be out of the city.

"This is Sunday and tomorrow a doctor is to look me over. Just now i am on home care. 2 meals per week with Meals on Wheels. 1 nurse 3 times per week, 1 homemaker 3 times per week, so you see i am getting the best of care."[2]

He missed the island. Nelson II remembers, "My father wanted us to rent a place and move to Victoria. I think he just wanted to be back on the island. He used to tell my wife Sue-Anne that if he started to feel better, he was going back to Copper Island and he was going to build her a house. But of course, that's the last place on earth she wanted to live." In any case, it was not to be. Nelson spent a year with his son and family and then was admitted to a veteran's hospital, St Vincent's in South Vancouver. He deteriorated quickly at that point and was there for about a month before he passed away on March 26, 1998, at the age of 89.

343197

DUNKIN — Nelson Edward, born January 22, 1909 in Olympia, Washington, died March 26, 1998 in Vancouver, B.C. Predeceased by his wife Mina on November 2, 1977 and granddaughter Arlene on July 9, 1981. Survived by son Nelson and wife Sue-Anne Dunkin of Vancouver and daughter Madge Vallee of Port Albern.; 10 grandchildren; and 5 great-grandchildren. A Funeral Service to be held in the chapel at Forest Lawn Funeral Home, 3789 Royal Oak Ave, Burnaby on Wednesday, April 1, at 2 p.m. No flowers by request.
FOREST LAWN FUNERAL HOME 299-7720

Debbie from Coastal Missions called to tell me the news of Nelson's passing into glory. A few days later, Nelson II contacted me to ask if I knew anyone familiar with the hymn "Be Thou My Vision," as his father had wanted it included in the funeral service. Since I had often taken my guitar to the old apartment in Coquitlam to sing the hymn to Nelson, I was glad to sing it at Nelson's funeral. The hymn has become increasingly popular in North America since that time, but I can take no credit for this; I am no great singer.

I travelled with some of the Coastal Missions crew to the funeral. Standing at the graveside at Forest Lawn Memorial Park in Burnaby, I found myself wishing I was not in the city but instead just coming up to the float on Copper Island with Nelson deftly navigating the logs to welcome me. Perhaps in the "new heavens and new earth," he will do just that. But on that grey day in the city, I felt far removed from the island that contained my memories of Nelson Dunkin.

4

Dave and Inger decided to have a memorial for Nelson on Copper Island, inviting the family and the people of Barkley Sound to honour him there. Inger said, "We were happy to have Madge and other Port Alberni community members come down on *Lady Rose* for the memorial service for Nelson that was held at the camp." Susan Scott, daughter of author R. Bruce Scott, attended as well. Madge told me she appreciated the memorial more than the funeral in Vancouver. "The one on Copper Island was a whole lot better. Because, I mean, he never wanted to live in the city; that was the last place he ever wanted to be. I guess there were the people who were at the camp, and Mary Scholey, and there were other people who came from Bamfield. I mean, they could relate much better to this memorial than a funeral in Vancouver."

"It was easy to love this unique child of God," said Ben. "He touched so many lives in a positive way. When he died, one person who knew him well told me, 'It's the end of an era.' Indeed, it seemed to be."

This cross carved by Nelson reminds us of the hope he maintained to the end of his days on earth.

---

[1] Letter to Leona Dolling, 1997

[2] Ibid

[3] "Obituaries," *The Province* and *Alberni Valley Times*, March 27, 1998

[4] findagrave.com/memorial/251673021/nelson-e-dunkin

50. The Legacy

After 29 years, I had the opportunity to return to Copper Island. I was running a Christian gap-year program for young adults. As a service project, I brought four students to Copper Island Camp in the spring of 2016 to help the staff ready the facilities for the summer. A few years before, I had met Aaron and Julie Otis—current Director of Ministries at the camp—and sent some young adults from another program, but didn't have the opportunity to accompany them. I quickly came to appreciate Aaron's gentle spirit, godly example and observable competence in leading ministries, both on Copper Island and throughout the year in Port Alberni.

I'm afraid I drove the camp staff crazy with my stories about Nelson Dunkin on that visit. I could hardly help myself—every time I turned around, I saw him there: in the view of the bay; in his house and the Dew Drop Inn, which were still standing; in the signs in the dining room that he carved; and in the very pebbles of Pebble Beach. The resident caretaker and his family were living on the bottom floor (with some additions upstairs) of Nelson's house, and the whole family soon became our friends. They took me up the stairs to what they called the "Heritage Room."

The upstairs suite where Nelson had once lived looked smaller than I remembered. Though it has the best view of the bay on the whole property, the camp set this room aside to honour and remember Nelson Dunkin and to be a place of quiet reflection. The old table and bench seat were still there, though the calendar on which he marked the days since Mina's passing was, of course, not on the wall. Blue carpet still lined his tiny bedroom, with the carvings of ships and whales attached. Recently, Madge donated a portrait of Nelson to the camp, and they hung it on the wall of this room to remind people of the former steward of Copper Island.

Please don't expect to have the same opportunity to visit Copper Island as I did. I have worked with summer camps all my life, and I know how they need to carefully guard their extraordinary properties. Copper Island Camp is not a holiday destination or a historical site—it is a refuge for children and workers to come away from the noisy world for a bit and hear the voice of God in the water, the trees, the people and the word of God. They will appreciate your prayers and—if so moved—your financial support, since the revenue from camper fees is a tiny fraction of what it costs to build, maintain and staff such a facility.

The author and his wife Sarah on Copper Island, Nelson's house in the background.

A few months after Nelson went home to be with Jesus, the *Alberni Valley Times* wrote an article about Dave and Inger, and about what they were accomplishing on Copper Island at the time:

> *Bible camp brings disadvantaged children to wilderness setting*
>
> *It's only 8 in the morning, but the 30 kids laughing and chasing one another around the decks of the* Lady Rose *are wide awake. The anticipation of arriving at the Copper Island Bible Camp has them in high spirits. Many of them have been looking forward to this day since leaving camp last summer. "I got up at 3 this morning," says 10-year-old Crystal Sampson, holding up three fingers to emphasize her point.*
>
> *She and several others came all the way from Saanichton this morning to catch the boat. The rest of her group came from reserves all over the island, including Port Alberni, Port Hardy, Nanaimo, and Duncan. "We want to provide native kids with an opportunity to learn about God in a wilderness setting," says Inger Logelin, director of the Copper Island Wilderness Camp. "It's an opportunity they might not otherwise get. I feel like there's a need for these kids to have a Christian camp experience."*
>
> *Logelin and her husband Dave have been running the camp for twelve years now. They have watched many kids grow up over the summers, and are now seeing their younger brothers and sisters come through. The camp has never advertised—they have never had to. They work through contacts on the reserves, and their reputation takes care of the rest. "We see so many good changes in kids' lives when you love them and respect them," says Logelin. "A lot of them come from troubled homes. It's good for them to be here with people who enjoy being with them."*
>
> *The staff on Copper Island are all there because they love it. Nobody, not even Inger and Dave, gets paid. The rewards are immeasurable, and the setting pristine. "It's a true wilderness*

camp," says Belle Ruiz, the Activities Director. "The property is over 100 acres, with a waterfall, and almost a mile of beach." The registration brochure boasts "an electronic and smoke-free zone!" What electricity there is comes from a waterwheel. In the late 1800s, the camp was the site of a trading post and copper mine. The abandoned mine shaft at the top of a hill is one of the camp's hiking destinations.

When the Logelins were given use of the property by Nelson Dunkin, who had lived there for decades previous, there was only brush, and a couple of old miner's shacks [well, you can't expect a newspaper article to be completely accurate!]. Since then, camp volunteers have built a mess hall and a two storey multi-purpose building, and have cleared the land for up to ten tents occupied by campers. "This was all float logs that we milled ourselves," says Logelin, entering the beautiful, high-ceilinged mess hall. "We didn't put a lot of money into these buildings, but we did put in a lot of sweat."

Dozens of kids and leaders will feast on home-style macaroni and cheese for this day's lunch. After lunch, there will be swimming, and tows behind the boat on a ski biscuit and body boards. The ocean is cold, and the sky overcast, but the kids are not daunted. Inger Logelin smiles at their enthusiasm. "They don't care what the weather's like."[1]

Others have since taken up the torch to lead the camp and provide the needed facilities. Almost every summer since 1987, campers from all over Vancouver Island—primarily First Nations—have come to Copper Island Camp. The current Director of Ministries, Aaron Otis, sent me a description of the continuing ministry of the Wilderness Retreat Society:

It is a privilege for Copper Island Bible Camp to have such a storied history. Our camp often reflects on Nelson's decision to sell the property for a loonie so local youth can hear the Good

News of Jesus Christ. It is a story I often share with campers, guests, and volunteers who come to this wonderful island location. Nelson Dunkin's life of faith and obedience to the Lord is an exemplary model for us to follow. It is a story that makes us pause and examine what is really important in life.

For many years, the good news was not seen as good news, as God's character and name were misrepresented through the abusive teachings of the residential schools. The Bible was misquoted and used to harm and manipulate First Nations people. As followers of Jesus, this should break our hearts and cause us to want to tell others who God really is.

At Copper Island Bible Camp, it's our calling and privilege to share who God is as revealed in Scripture. "Created for a Purpose," "Living your Purpose" and "Finding your Purpose" are themes we cycle through every summer. Don't we all need to hear these truths that God loves us, knows us and has a plan for each one of us?

Every camper is treated with dignity, love, and respect. The walls we all build to protect ourselves come down quickly in the peaceful setting of the island and the genuine offer of friendship. Over the years, many youth have heard and responded to the good news and begun a relationship with the Creator Jesus.

As a year-round ministry, our staff also serves in various communities on Vancouver Island, mentoring, tutoring and encouraging youth and young adults. A joy for us is to see youth grow in their relationship with God and return to Copper Island to volunteer and work with us as staff. It's a lot of work planning for camps and caring for the facilities throughout the year, but the work is sweet as we remember we are a part of something bigger than ourselves.

Our hope and prayer is much like that of Nelson and Mina Dunkin: that God will continue to use Copper Island as a place of blessing where people can find hope, purpose and a friend who will never leave or forsake them.

- Aaron Otis

Bill Priest told me, "I got back to Barkley Sound in 2021, and I stayed with Jim and Dodi over in Kildonan. He gave me a tin boat to putter about with for part of the day, and I ended up over at Copper Island. I just went up to the entrance, and didn't go in to talk to people, and didn't want to spend too much time there. Kind of told my story and bawled my eyes out. I was glad the property didn't get lost to private development and is being used for the purpose that Nelson had intended."

The legacy of Nelson and Mina Dunkin is more than a parcel of land on an island in Barkley Sound—that is just its lovely venue. Nelson and Mina invested in the lives of people and, ultimately, in the kingdom of God. The return on their investment is not a loonie on a piece of string; it is the women, men, youth and children whose character and destiny were gently altered through a pair of faithful and generous stewards of what God entrusted to them. May their number multiply for generations to come.

Nelson E. Dunkin

[1] Kirstin Abercrombie, "Bible camp brings disadvantaged children to wilderness setting," *Alberni Valley Times*, August 13, 1998

Sources

Interviews (in person unless otherwise indicated):

- Brian Burkholder – *24-Feb-23*
- Aaron Otis – *15-Mar-23 (phone)*
- Ruth Sadler – *25-Mar-23 (phone)*
- Harold Sadler – *24-Mar-23 (phone)*
- Tom & Debbie Maxie – *27-Mar-23*
- Heather Arnott – *10-Apr-23*
- Dave & Inger Logelin – *18-Apr-23 (email)*
- Ben Potter – *23-Apr-23*
- Madge Vallee – *23-Apr-23*
- Patty Cameron – *24-Apr-23*
- Wendall Ferrell – *24-Apr-23*
- Earl Johnson – *25-Apr-23*
- Bill Irving – *10-May-23 (phone)*
- Joan (Petunia) Getman – *11-May-23*
- Roy Getman – *11-May-23*
- Nelson Dunkin II – *15-May-23 (phone)*
- Margaret Stewart – *05-Jun-23 (video)*
- Pat Rafuse – *11-Jun-23*
- Bill Priest – *12-Jun-23 (video)*
- Peter Horton – *13-Jun-23 (phone)*
- Rich Parlee – *27-Aug-23 (phone)*
- Leona Dolling – *19-Sep-23 (phone)*
- Rick Charles – *06-Nov-23 (phone)*
- Len Gedak – *07-Jan-24 (phone)*

Books:

- Cathy Converse and Beth Hill, *The Remarkable World of*

Frances Barkley: 1769-1845, 2008

- Rev Chas. Moser, *Reminiscences of the West Coast of Vancouver Island*, 1926
- Earl Johnson, *Looking Astern*, 2018
- Jan Peterson, *Journeys Down the Alberni Canal to Barkley Sound*, 1999
- John T Walbran, *British Columbia Coast Names, 1592-1906: their origin and history*, 1909
- Kathryn Bridge & Kevin Neary, *Voices of the Elders, Huu-ay-aht Histories and Legends*, 2013
- Pat Garcia, *Bamfield: Looking Back*, 2010
- R. Bruce Scott, *People of the Southwest Coast of Vancouver Island*, 1974

Newspaper, Web and Magazine Articles:

- Adele Wickett, "Remembering Nelson Dunkin," *Island Christian Info*, June 1998
- Bob Miller, "Nelson Dunkin," *Black-and-White Album Blog*
- "Centennial Park Sign," *Alberni Valley Times*, August 7, 1968
- "Copper Island resident buried locally," *Alberni Valley Times*, November 15, 1977
- "Guides & Brownies," *Alberni Valley Times*, October 12, 1973
- Katie Poole, "An Island of Solitude," *Tri-City News*, November 17, 1991
- Kirstin Abercrombie, "Bible camp brings disadvantaged children to wilderness setting," *Alberni Valley Times*, August 13, 1998
- "The Mission of the Shantymen," *Life Magazine*, January 11, 1954

- R. Bruce Scott, "Independent People," *The Daily Colonist*, February 9, 1969
- "Something about Mary," *Times Colonist*, Mar 25, 2001
- "War Brides Greeted by B.C. Kinfolk," *The Vancouver Sun*, October 13, 1944

Historical Records:

- "Census of the United States, Population Schedule," 1910, 1920, 1930 - myheritage.com
- "Nelson Edward Dunkin," "Jamesina Dunkin (born MacIver)," "Jonathan Cook Dunkin," "Rhoda May Dunkin (born Tedham," "Mary MacIver," "Murdo MacIver" - myheritage.com
- Thank you to the Bamfield Historical Society and Alberni Valley Museum Archives for your kind assistance.

Letters (written by Nelson to:)

- Heather Arnott
- Jim & Sarah Badke
- Leona Dolling
- Pat Rafuse
- Patty & Don Cameron
- Ron Pollock

Photographs (supplied but not necessarily taken by:)

- Aaron Otis
- Alberni Valley Museum
- Bamfield Community Museum and Archives
- Dave & Inger Logelin
- Heather Arnott
- Jim Badke

- Len Gedak
- Leona Dolling
- Lynn Starter
- Madge Vallee
- Margaret Stewart
- Pat Rafuse
- Patty Cameron
- Peter Horton
- Phil Hood
- Tom & Debbie Maxie

Artwork

- Jay Bradley
- Lynn Starter
- Rick Charles
- Roy Getman

Carvings and Projects (shown/photographed by:)

- Aaron Otis
- Brian Burkholder
- Dave & Inger Logelin
- Heather Arnott
- Heather Cooper
- Leona Dolling
- Madge Vallee
- Pat Rafuse
- Patty Cameron
- Wendall Ferrell

Also by Jim Badke:

As The Eagle Walks

If the Apostle John was still living on the island of Patmos, would you want to visit him?

The Mummy of Fisher Creek

A young couple has found a pair of boots in a cave—and someone is still inside them.

Project VI

A research visa to an affluent offshore country near the West Coast turns into a nightmare.

Camp Liverwurst

A series for middle-school kids set at a camp that is out of this world and a whole lot of fun.

---Available on Amazon---

Manufactured by Amazon.ca
Bolton, ON